THE NATIONAL ACADEMIES
Advisers to the Nation on Science, Engineering, and Medicine

The **National Academy of Sciences** is a private, nonprofit, self-perpetuating society of distinguished scholars engaged in scientific and engineering research, dedicated to the furtherance of science and technology and to their use for the general welfare. Upon the authority of the charter granted to it by the Congress in 1863, the Academy has a mandate that requires it to advise the federal government on scientific and technical matters. Dr. Ralph J. Cicerone is president of the National Academy of Sciences.

The **National Academy of Engineering** was established in 1964, under the charter of the National Academy of Sciences, as a parallel organization of outstanding engineers. It is autonomous in its administration and in the selection of its members, sharing with the National Academy of Sciences the responsibility for advising the federal government. The National Academy of Engineering also sponsors engineering programs aimed at meeting national needs, encourages education and research, and recognizes the superior achievements of engineers. Dr. C. D. Mote, Jr., is president of the National Academy of Engineering.

The **Institute of Medicine** was established in 1970 by the National Academy of Sciences to secure the services of eminent members of appropriate professions in the examination of policy matters pertaining to the health of the public. The Institute acts under the responsibility given to the National Academy of Sciences by its congressional charter to be an adviser to the federal government and, upon its own initiative, to identify issues of medical care, research, and education. Dr. Harvey V. Fineberg is president of the Institute of Medicine.

The **National Research Council** was organized by the National Academy of Sciences in 1916 to associate the broad community of science and technology with the Academy's purposes of furthering knowledge and advising the federal government. Functioning in accordance with general policies determined by the Academy, the Council has become the principal operating agency of both the National Academy of Sciences and the National Academy of Engineering in providing services to the government, the public, and the scientific and engineering communities. The Council is administered jointly by both Academies and the Institute of Medicine. Dr. Ralph J. Cicerone and Dr. C. D. Mote, Jr., are chair and vice chair, respectively, of the National Research Council.

www.national-academies.org

THE ARC

OF THE ACADEMIC RESEARCH CAREER

Issues and Implications for U.S. Science and Engineering Leadership

Summary of a Workshop

Beryl Lieff Benderly, Rapporteur

Committee on Science, Engineering, and Public Policy

Policy and Global Affairs

NATIONAL ACADEMY OF SCIENCES,
NATIONAL ACADEMY OF ENGINEERING, AND
INSTITUTE OF MEDICINE
OF THE NATIONAL ACADEMIES

THE NATIONAL ACADEMIES PRESS
Washington, D.C.
www.nap.edu

THE NATIONAL ACADEMIES PRESS   500 Fifth Street, NW   Washington, DC 20001

NOTICE: The project that is the subject of this report was approved by the Governing Board of the National Research Council, whose members are drawn from the councils of the National Academy of Sciences, the National Academy of Engineering, and the Institute of Medicine. The members of the committee responsible for the report were chosen for their special competences and with regard for appropriate balance.

This study was supported by Grant No. B2013-31 between the National Academy of Sciences and the Alfred P. Sloan Foundation. Any opinions, findings, conclusions, or recommendations expressed in this publication are those of the authors and do not necessarily reflect the views of the organization that provided support for the project.

International Standard Book Number 13: 978- 0-309-29896-4
International Standard Book Number 10: 0-309-29896-2

Additional copies of this report are available from the National Academies Press, 500 Fifth Street, NW, Room 360, Washington, DC 20001; (800) 624-6242 or (202) 334-3313;   http://www.nap.edu

Copyright 2014 by the National Academy of Sciences. All rights reserved.

Printed in the United States of America

PLANNING COMMITTEE FOR THE WORKSHOP ON THE ARC OF THE ACADEMIC RESEARCH CAREER

RICHARD N. ZARE [NAS] (*chair*), Marguerite Blake Wilbur Professor, Stanford University

DAVID DANIEL [NAE], President, The University of Texas at Dallas

DIANE E. GRIFFIN [NAS, IOM], Alfred and Jill Sommer Professor, Chair in Molecular Microbiology and Immunology, Johns Hopkins Bloomberg School of Public Health

SHIRLEY M. MALCOM [NAS] Head of the Directorate for Education and Human Resources Programs, American Association for the Advancement of Science

PERCY A. PIERRE [NAE], Vice President and Professor Emeritus, Michigan State University

E. ALBERT REECE [IOM], Vice President for Medical Affairs, Bowers Distinguished Professor and Dean, School of Medicine, University of Maryland, Baltimore

Staff
KEVIN FINNERAN, Director
MARIA LUND DAHLBERG, Research Associate
NEERAJ GORKHALY, Research Associate
MARION RAMSEY, Administrative Associate (until October 2013)

Consultants
BERYL LIEFF BENDERLY, Consultant Writer and Rapporteur

COMMITTEE ON SCIENCE, ENGINEERING, AND PUBLIC POLICY

RICHARD N. ZARE [NAS] (*chair*), Marguerite Blake Wilbur Professor, Stanford University

LINDA M. ABRIOLA [NAE], Dean of Engineering, Tufts University

SUSAN ATHEY [NAS], Professor, Graduate School of Business, Stanford University

MOSES H. W. CHAN [NAS], Evan Pugh Professor of Physics, Pennsylvania State University

RALPH J. CICERONE [NAS] (*ex-officio*), President, National Academy of Sciences

PAUL CITRON [NAE], Vice President (retired), Technology Policy and Academic Relations, Medtronic, Inc.

DAVID DANIEL [NAE], President, The University of Texas at Dallas

GORDON R. ENGLAND [NAE], President, E6 Partners LLC

HARVEY V. FINEBERG [IOM] (*ex-officio*), President, Institute of Medicine

DIANE E. GRIFFIN [NAS, IOM], Alfred and Jill Sommer Professor, Chair in Molecular Microbiology and Immunology, Johns Hopkins Bloomberg School of Public Health

JOHN G. HILDEBRAND [NAS], Regents Professor, Department of Neuroscience, University of Arizona

DAVID KORN [IOM], Consultant in Pathology, Massachusetts General Hospital; Professor of Pathology, Harvard Medical School

C. D. MOTE, JR. [NAE] (*ex-officio*), President, National Academy of Engineering

PERCY A. PIERRE [NAE], Vice President and Professor Emeritus, Michigan State University

E. ALBERT REECE [IOM], Vice President for Medical Affairs, Bowers Distinguished Professor and Dean, School of Medicine, University of Maryland, Baltimore

MICHAEL S. TURNER [NAS], Rauner Distinguished Service Professor, Kavli Institute for Cosmological Physics, The University of Chicago

NANCY S. WEXLER [IOM], Higgins Professor of Neuropsychology, Colleges of Physicians and Surgeons, Columbia University

PETER WOLYNES [NAS], D.R. Bullard-Welch Foundation Professor of Chemistry, Center for Theoretical Biological Physics-BCR, Rice University

Staff
KEVIN FINNERAN, Director
TOM ARRISON, Senior Program Officer
GURU MADHAVAN, Program Officer
MARIA LUND DAHLBERG, Research Associate
NEERAJ GORKHALY, Research Associate (until February 2014)
MARION RAMSEY, Administrative Associate (until October 2013)

Reviewer Acknowledgments

This report has been reviewed in draft form by individuals chosen for their diverse perspectives and technical expertise, in accordance with procedures approved by the National Academies' Report Review Committee. The purpose of this independent review is to provide candid and critical comments that will assist the institution in making its published report as sound as possible and to ensure that the report meets institutional standards for quality and objectivity. The review comments and draft manuscript remain confidential to protect the integrity of the process.

We wish to thank the following individuals for their review of this report: Susan Carlson, University of California; Paul Citron, University of California-San Diego; Donna Ginther, Kansas University; Peter McPherson, Association of Public and Land-Grant Universities; and Henry Sauermann, Georgia Institute of Technology.

Although the reviewers listed above have provided many constructive comments and suggestions, they were not asked to endorse the content of the report, nor did they see the final draft before its release. Responsibility for the final content of this report rests entirely with the rapporteur and the institution.

Contents

1 INTRODUCTION	**1**
2 BACKGROUND TO CHANGE	**5**
Structure of the Academic Research Career	5
The Unstable Arc	7
3 GETTING STARTED	**13**
The Economics of Early Careers	15
Aspirations and Realities	19
Family Matters	22
4 THE TENURE TRACK AND BEYOND	**27**
It's All in the Timing	29
After Tenure	33
5 MOVING INTO RETIREMENT	**37**
Incentives, not Coercion	40
Remaining Relevant	41
Untapped Assets	44
6 THE OTHER ACADEME	**45**
7 LOOKING AHEAD	**49**
APPENDIXES	**53**
Appendix A: Workshop Agenda	55
Appendix B: Speakers' Biographies	59

1
Introduction

    America's research universities have undergone striking change in recent decades, as have many aspects of the society that surrounds them. This change has important implications for the heart of every university: the faculty. To sustain their high level of intellectual excellence and their success in preparing young people for the various roles they will play in society, universities need to be aware of how evolving conditions affect their ability to attract the most qualified people and to maximize their effectiveness as teachers and researchers. This workshop summary addresses the challenges universities face from nurturing the talent of future faculty members to managing their progress through all the stages of their careers to finding the best use of their skills as their work winds down.
    Gender roles, family life, the demographic makeup of the nation and the faculty, and the economic stability of higher education all have shifted dramatically over the past generation. In addition, strong current trends in technology, funding, and demographics suggest that change will continue and perhaps even accelerate in academe in the years to come. Among the forces now propelling America's universities toward an uncertain future are: increasing financial pressures on institutions, the research enterprise, and students; the advent of computer-based instruction on a worldwide scale; and the growing internationalization of both higher education and research.
    One central element of academic life has remained essentially unchanged for generations, however: the formal structure of the professorial career. Developed in the mid-nineteenth and early twentieth centuries to suit circumstances quite different from today's, and based on traditions going back even earlier, this customary career path is now a source of strain for both the individuals pursuing it and the institutions where they work.
    Universities' effectiveness in supporting the careers of their scientific faculty matters, because faculty members pursuing that traditional career path at

research universities play a crucial role in the nation's research enterprise. Collectively, tenured and tenure-track faculty researchers account for much of the scientific and technological progress that underlies the nation's prosperity, security, well-being, and world leadership. For generations, secure academic positions have given faculty members the stability and resources to pursue their work. Only by assuring that gifted and highly qualified individuals from a wide variety of backgrounds are able to enter and thrive in scientific and technical research careers at academic institutions can this vital progress and leadership continue.

Changing conditions have inspired a number of universities to develop innovative approaches that attempt to adapt long-familiar practices, procedures, and concepts designed for different times to the challenges facing today's faculty members and institutions. In developing these new systems, it is important that universities heed how changes to policies governing one stage of a career can have repercussions for the other stages.

The Committee on Science, Engineering, and Public Policy organized the workshop summarized in this report to examine major points of strain in academic research careers from the point of view of both the faculty members and the institutions. Although the issues discussed are relevant to faculty in all disciplines, the focus in this workshop was on the biological sciences, physical sciences, the social sciences, and engineering. The workshop was held in Washington, DC, in the National Academy of Sciences Building on September 9th and 10th, 2013. The gathering brought together national experts from a variety of disciplines and institutions to highlight practices and strategies already in use on various campuses and to identify issues as yet not effectively addressed. Although the workshop was designed to study current conditions and future possibilities, it was not intended to make policy recommendations. It comprised six sessions spread over a day and a half. The first day spanned the academic career arc with sessions entitled "Overview of Challenges to U.S. Universities and Academic Science and Engineering Careers," "Getting Started: Early Career Bottleneck," "The Family v. The Workplace: Mid-Career Priorities," and "Beautiful Sunsets: A Fulfilling Late-career Transition." The workshop's second day continued with the sixth session, "Reports from the Field: Examples of Innovative Approaches," and concluded with the final section, "Opportunities for Action."[1]

This report aims to summarize the issues and information presented and discussed at the workshop. It is not, however, a chronological account of the gathering. Instead, it is organized around the broad themes covered in the presentations and discussions. Ideas, data, and participants' statements are introduced where they are relevant to a topic, not necessarily when they appeared in the program.

---

[1] A full agenda for the workshop can be found in Appendix A; biographies of the speakers can be found in Appendix B.
[2] Many universities also have research faculty positions that might enable independent research but

INTRODUCTION

The report begins with background information that provides a framework for the discussion. Next, it covers the major phases of the academic career, highlighting the stress points that connect them. These phases are entry into academe, the tenure decision and the mid-career years that follow it, and the transition to retirement. The report then considers important issues outside that traditional ladder involving the many academics in the burgeoning ranks of "off-track" or "non-ladder" faculty. The report closes with workshop participants' observations on opportunities for future action.

The report has been prepared by the workshop rapporteur as a factual summary of what occurred at the workshop. The planning committee's role was limited to setting the agenda and convening the workshop. The views contained in the report are those of individual workshop participants and do not necessarily represent the views of all workshop participants, the planning committee, or the National Research Council.

2
Background to Change

In the course of addressing specific aspects of an academic career, all of the speakers contributed to a description of the overall structure of an academic research career. Before drilling down into the details, it is useful to sketch this general framework and see the big picture.

STRUCTURE OF THE ACADEMIC RESEARCH CAREER

Implicit in all the workshop discussions was a shared understanding of how a traditional academic career progresses. Basic to this system is tenure, a promise of lifetime employment as a faculty member at a particular institution. Once earned, tenure is very rarely lost, and then only for severe misbehavior by the individual or extreme financial difficulty of the institution. Attaining tenure makes an individual a participant in certain aspects of the institution's governance, including decisions about hiring and granting tenure to others.

Only particular job slots that the institution designates as belonging to the so-called tenure track carry the possibility of providing tenure. Positions not on the tenure track may involve similar duties, such as teaching, research, advising students, and serving on certain committees, but they do not offer a promise of permanence or provide the status within the academic community that tenure does. In former times, tenure constituted the institution's promise to provide all or at least part of the faculty member's salary until he or she elected to leave or reached retirement age. In many institutions today, however, tenure often provides only a platform from which faculty members can engage in the competition for funding to support their work and provide their own salaries. The extent of dependence on outside support differs considerably across disciplines.

To have a chance of winning tenure, a scholar must first secure a position that is explicitly designated as part of the tenure track. Positions in this sequence carry one of three ranks: assistant professor, which is the lowest;

associate professor; and full professor, which is the highest. Tenure is not automatic and successful candidates must win the approval of both departmental colleagues and the higher administration of their institution. Assistant professors serve a probationary period that can last up to 7, or sometimes 10, years. During this time, the candidate strives to amass a record of research publications and successful grant applications—and, to a much lesser extent, of teaching and service to department and university—that the institution deems acceptable. Unsuccessful candidates must leave the university and seek employment elsewhere. The years preceding tenure therefore constitute a make-or-break period of great tension, long hours, and hard work.

Appointment to a tenure-track assistant professorship certifies the individual as an independent investigator who has the institution's backing in the competition for research grants.[2] In fields that require laboratories, equipment, and workers to produce research results, universities provide new assistant professors start-up money to establish their laboratories. In return, the new assistant professor is expected to start providing money to support the laboratory within a few years by winning competitive grants, usually from the federal government.

The decision about whether to grant tenure comes after a predetermined number of years, with a symbolic "tenure clock" marking the time to one of the most fateful moments in the assistant professor's life. Attaining tenure generally coincides with promotion to the rank of associate professor, which brings higher pay and greater status and recognition. Above that, only the rank of full professor remains, although within that rank, many universities award additional recognition in the form of distinguished, University, or named professorships. Reaching that highest rank places a scholar among the senior faculty members of the institution and, for many, indicates attainment of a successful career. This final promotion decision is again made by the candidate's departmental colleagues and the higher administration, and again, the decision overwhelmingly depends on their estimation of the quantity and quality of the candidate's research. The requirement to do top-tier research is paramount at doctorate-granting, research-intensive universities.

Attaining a full professorship has no set time limit, and some faculty members never reach that rank and remain associate professors throughout their careers. At retirement, both full and associate professors may receive emeritus status, a largely honorific title that indicates a continuing connection to the department and may, depending on the university's resources, include such perquisites as use of an office and lab, computer accounts, administrative support, and the like.

Given the stringent requirements for advancement at middle- and upper-tier institutions, competition marks most stages of a topflight academic career—competition to be hired into a tenure-track position, to win funding, to make

---

[2] Many universities also have research faculty positions that might enable independent research but that typically do not provide access to tenure and are considered less prestigious.

discoveries that will make one's name, to be first to publish them. Although this competition for top people occurs in all fields, the need to attract external research funding is particularly important in the sciences and engineering. Particularly successful competitors—those who consistently receive substantial research support and achieve publications in prestigious journals and build eminent reputations within their fields—can win prestigious and lucrative honors and receive appealing offers to move to other institutions, bringing their productive labs and attendant grants with them to their new academic homes. Leading researchers in a field may make several such moves.

THE UNSTABLE ARC

In the early 1970s, as a graduate student looking forward to her future, Shirley Malcom, head of the Directorate for Education and Human Resources Programs, American Association for the Advancement of Science, saw "an expected arc to my life and my career," she told the workshop in its opening session. She and her fellow Ph.D. candidates, she said, were led to believe that a new Ph.D.'s career would follow the same course as their professors'. "When I finished my Ph.D.," Malcom recalled thinking, "I would enter a tenure-track position...I would gain research independence [and] get an early first grant,...would get tenure,...would be promoted" and would remain a member of a university faculty until she reached the mandatory retirement age of 70. Deviating from the "pathway [that] was set out" by taking a position outside of academe, she said, constituted "choosing an alternative career."

Malcom, an African American woman, recalls believing this even though the research university professoriate she hoped to join consisted of people "quite different" from herself: overwhelmingly male and white, with women constituting only 2 percent of the full professorships in science and engineering and minority group members hardly visible in those ranks at all.

From the apparent predictability and stability that students saw four decades ago, "we have moved...to a time of flux and uncertainty in higher education," Malcom continued. Today's graduate school students and aspiring scientists, who are the "academic progeny" of Malcom's generation, need a different set of expectations for the very different world of today and tomorrow, she said. This generation "will likely not have [their] mentor's career" and should not "even expect it, because things really have changed a lot."

Change has been so great, she continued, that for today's graduate students and young Ph.D.s, the traditional progression from graduate student to tenured professorship is now, statistically at least, the "alternative career." The number of tenure-track positions has increased very little in recent decades, except in engineering, but the number of Ph.D.s awarded in science, technology, engineering, and mathematics (STEM) fields has increased very rapidly, far outstripping the availability of tenure-track positions. Edie Goldenberg, professor of political science and public policy at the University of Michigan, and Henry Sauermann, assistant professor of strategic management at the Ernest

Scheller, Jr., College of Business at Georgia Tech, observed that, as the academic job market has become increasingly overcrowded new Ph.D.s in many fields who hope for academic careers now must spend up to 5 or more years as postdoctoral researchers (known as postdocs), gaining training needed to advance their careers. Only then can they even attempt to look for a tenure-track position, added Mary Ann Mason, co-director of the Center for Economics & Family Security at the University of California-Berkeley School of Law, and the odds of success have steadily declined.

The minority who manage to secure a foothold on the tenure track are in their late thirties to early forties, on average, before they win their first independent grant, which establishes a scientist as an independent principal investigator (PI), Malcom noted. The route to a faculty career has become so protracted, she said, that since 2002 the percentage of PIs older than 61 has exceeded the percentage under age 36, even though the early years of scientific careers are often thought to be the most creative.

A second great change has been the composition of both the graduate student and postdoc populations and the professoriate, as women have entered graduate school and the academic profession in large numbers. Robert Hauser, the executive director of the Division of Behavioral and Social Sciences and Education at the National Research Council and the former director of the Center for Demography of Health and Aging at the University of Wisconsin, who joined Malcom in setting the stage for the issues the workshop would consider, noted that the percentage of women earning STEM doctorates has risen substantially in all fields, but especially in the life sciences. In 2009, women earned 55.6 percent of Ph.D.s in the life sciences, and about 30 percent in the physical sciences, mathematics, and engineering [3]

Women have also made striking gains on science faculties, Malcom noted. In 1993, they constituted 8 percent of the full professors in the sciences, engineering, and health fields at research universities and just under 10 percent in all institutions. By 2010, those figures stood at just over 20 percent and nearly 25 percent, respectively (see Figure 2-1). The percentage of full professors in those fields who were members of an underrepresented minority, however, remained below 5 percent at research universities and about 5 percent at all institutions in 2010 (see Figure 2-2).

Third, the economic robustness of research universities, and of the scientific, scholarly, and educational activities that they support, has noticeably deteriorated. As highlighted by the 2012 National Research Council report on research universities,[4] major causes include "unstable federal funding,...an erosion of state funding for higher [education], a dismantling of the large corporate labs by business and industry, [and inadequate] management and [in]efficiency," Malcom said.

---

[3] Table 335 *Doctorate Recipients From U.S. Universities: Summary Report 2008–09,*
[4] Research Universities and the Future of America: Ten Breakthrough Actions Vital to Our Nation's Prosperity and Security. (Washington, DC: The National Academies Press, 2012).

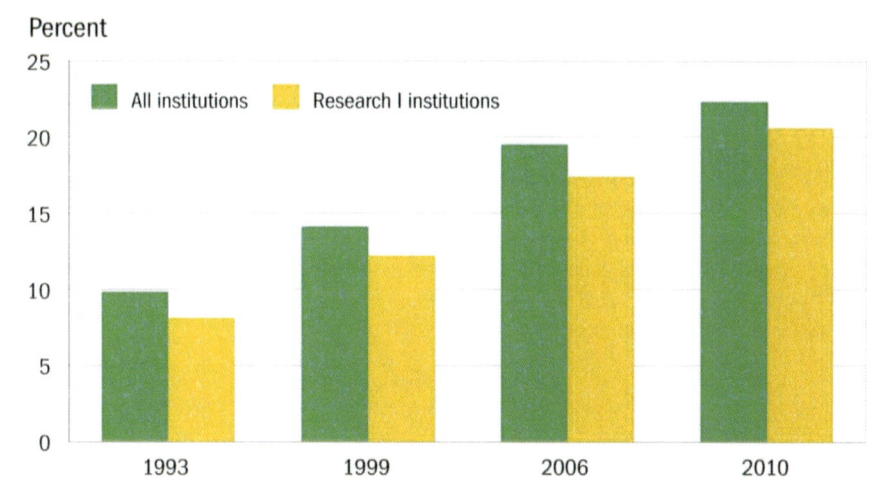

Figure 2-1 Women as a percentage of full-time, full professors with science, engineering, and health doctorates, by institution of employment: 1993–2010.
SOURCE: Women, Minorities, and Persons with Disabilities in Science and Engineering: 2013; www.nsf.gov/statis.

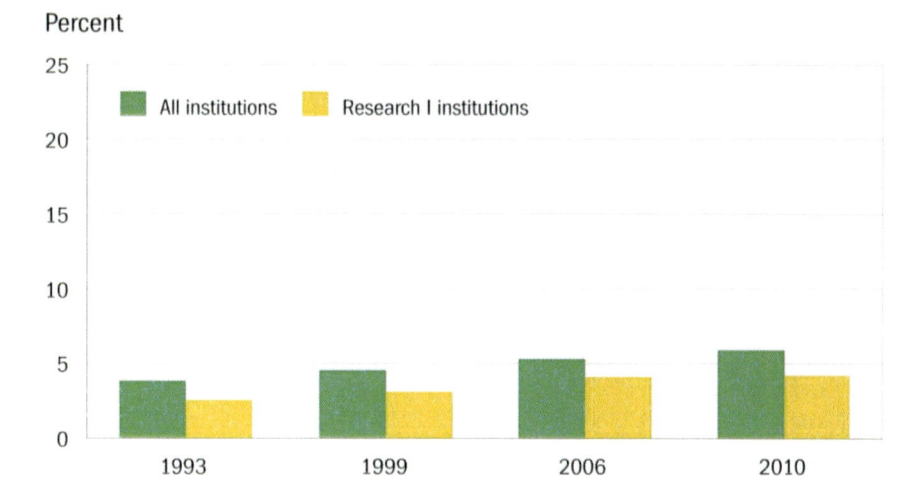

Figure 2-2 Underrepresented minorities as a percentage of full-time, full professors with science, engineering, and health doctorates, by institution of employment: 1993–2010.
SOURCE: Women, Minorities, and Persons with Disabilities in Science and Engineering: 2013; www.nsf.gov/statistics/wmpd/.

One response to these pressures has been a striking increase in the use of contingent or adjunct faculty, who cost less than tenure-track professors and do not require the start-up packages and lifelong job commitment that tenured faculty receive. The "adjunctification" of university faculties is "a huge change in the nature of academic employment," Hauser said. Full-time tenured faculty at all U.S. postsecondary institutions declined from 29 to 16.7 percent of instructional staff between 1975 and 2011 and tenure-track faculty, from 16 to 7 percent. Meanwhile "huge increases occurred in full-time non-tenure-track faculty [from 10.3 to 15.4 percent] and especially in part-time faculty," from 24 to 41 percent, he said (see Figure 2-3).[5] "That is a big change in the character of the academic workforce over these years."

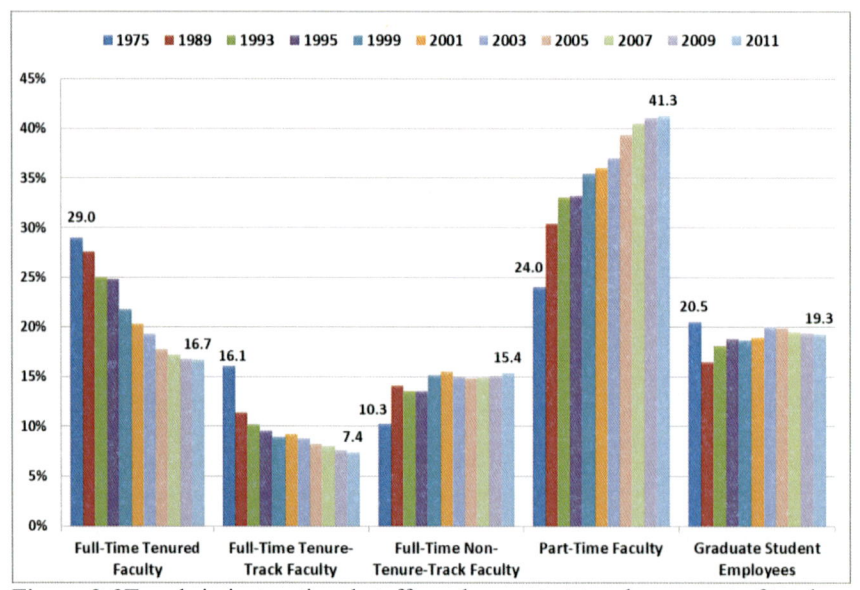

Figure 2-3 Trends in instructional staff employment status, by percent of total instructional staff for all institutions, national totals: 1975–2011.
NOTES: Figures for 2011 are estimated. Figures from 2005 have been corrected from those published in 2012. Figures are for degree-granting institutions only, but the precise category of institutions included has changed over time. Graduate student employee figures for 1975 are from 1976. Percentages may not add to 100 due to rounding.
SOURCE: U.S. Department of Education, IPEDS Fall Staff Survey. Tabulation by the American Association of University Professors (AAUP) Research Office, Washington, D.C. Released April 2013.

---

[5] Although this trend is more pronounced in non-research institutions, it is common at research universities as well.

Other important changes now affecting academic careers include the abolition, as of 1994, of mandatory retirement for tenured faculty; the rapid growth of 2-year colleges as a major source of undergraduate instruction; and the advent of online and blended instruction, which carries as yet unknown but potentially significant effects on existing academic institutions.

Together, all these changes have introduced major new issues to campus life. While Malcom's professors mostly had wives at home to raise their children and run their households, today both female and male faculty members must balance the demands of academic careers and family life. Perhaps most crucially for career success, the current elongated training period and pretenure years overlap women's prime years of fertility and both genders' major time for family formation. Today's faculty members—and their employers—must also negotiate the issues surrounding retirement, which has become discretionary rather than mandatory at a specified age.

The seemingly stable career arc that appeared to lie before aspiring academics during Malcom's graduate school days has irrevocably shattered. The more complex paths that today's and tomorrow's academics must follow, and ways that institutions may help or hinder their journeys, are the topics of the remaining sessions.

3
Getting Started

For young people aspiring to academic careers, the altered academic scene that Shirley Malcom and Robert Hauser described forms the context of their efforts to launch their professional lives as scientists. Amid the changes in the composition and size of the cohorts seeking to become faculty members and the increasing economic instability of both universities and research funding, something else besides the structure of the academic career has remained essentially constant: the number of faculty positions offering the possibility of tenure.

A mix of permanence and change has therefore produced "an excess supply of Ph.D.s and very uncertain demand [for their services in academe], given the challenges facing the contemporary university," said Donna Ginther, a professor of economics at the University of Kansas, introducing the discussion of the early years of the traditional academic arc. Of course, many Ph.D.s intend to pursue careers outside of academia, but the purpose of the workshop was to look at the academic career path.

The challenges facing young people aspiring to follow that path have become increasingly demanding and their chances of attaining their goal increasingly diminished. As Ginther and Malcom suggested, however, universities continue to admit graduate students and recruit postdocs based not on the career opportunities awaiting them but on the supply of grant funding and on professors' and departments' resulting need for their labor in laboratories and classrooms.

Today, competition is most intense, at least numerically speaking, at the earliest stages of the career. Four or five decades ago, new Ph.D.s who studied under well-recognized professors had a reasonable assurance of landing an assistant professorship if they did well in their studies. Today, far more new Ph.D.s compete for a pool of assistant professorships that has grown much more slowly, and the chances of reaching that crucial first toehold on the tenure ladder have shrunk. This is one reason why many young scientists who cannot launch independent academic careers accept postdoctoral positions as researchers in professors' labs.

Ostensibly providing a year or two of additional training to equip the new Ph.D.s to better compete for an assistant professorship, Ginther noted that these appointments in fact more often become low-paid academic jobs supported by soft money that can last 5 years or more. In practical terms, postdocs are "the workhorses of the research community," said Mary Ann Mason. Providing cheap, highly skilled labor paid for out of professors' grants, postdocs were 43 percent of the first authors of papers in *Science*, according to a 1999 study, she noted (see Box 3-1).

Box 3-1
Results from the Sigma Xi Postdoctoral survey.

"Postdocs perform a substantial fraction of the skilled work in research labs and are responsible for a disproportionate share of the new discoveries. **A 1999 study found that 43 percent of first authors of research article in *Science* were postdocs."**

- Geoff Davis, author of the Sigma Xi Postdoctoral Survey

SOURCE: NSF-NIH Survey of Graduate Students and Postdocs in Science and Engineering, (2008).; Davis, G. Improving the Postdoctoral Experience: An Empirical Approach. In R. Freeman and G. Davis (Eds.). The Science and Engineering Workforce in the United States. Chicago, IL. NBER/University of Chicago Press. (2006).

Because of rising production of Ph.D.s and stagnant numbers of professorships, the number of postdocs has grown rapidly over the years. As Joan Girgus, professor of psychology and special assistant to the dean of the faculty on gender equity at Princeton University, noted, the postdoc population at Princeton University has doubled in a decade, for example. Indeed, as several participants noted, no one knows exactly how many postdocs there are in the country and most published figures are underestimates. Girgus noted that although Princeton limits the term of a postdoctoral appointment to three years in most fields and five years in biomedicine, many of these postdocs then move into research positions that are essentially the same but are given a different title. She added that "We do not know how many of those there are."

As generally poorly paid temporary workers, postdocs have a very ambiguous status on most campuses, neither students nor faculty. "My sense about postdocs everywhere is that they are a pretty much forgotten population," said Goldenberg. For most of these "trainees," the appointment becomes either a prelude to further ambiguous soft-money academic positions or a jumping-off place for seeking work outside of academe, be it in industry, government, or elsewhere. The academic world has long regarded such work as "alternative careers," and most graduate programs do not provide adequate preparation. As

Malcom noted, however, the so-called alternative is now the norm for the majority of Ph.D.s in most scientific fields.

THE ECONOMICS OF EARLY CAREERS

Ginther began the analysis by stressing "that each academic field is a separate labor market. You can't combine all of science together and say, 'This is what is happening.' You can't combine all academics together and say, 'This is what is happening.' " Salary rates differ among academic fields "because there are different demands for our services, both inside and outside of academia." Faculty members who have real opportunities to earn good incomes off campus, such as physicians, lawyers, economists, and some scientists, tend to command higher incomes on campus, too.

Ginther argued that science breaks down into four broad categories: life science, including biomedical, agricultural, and environmental sciences; physical science, including chemistry and physics, geosciences, and math and computer sciences; social science, including psychology, sociology, and economics; and engineering. These broad disciplinary categories, however, all contain large numbers of specialized fields and subfields, each of which, to some extent, constitutes a distinct labor market.

Three of these four categories have seen significant growth in the number of Ph.D.s produced in recent decades, Ginther noted. Between 1980 and 2010, annual production doubled in the life and physical sciences, from under 6,000 to almost 12,000 and from 4,000 to 8,000, respectively. Engineering Ph.D. production grew from 2,000 in 1980 to 8,000 in 2010. Growth was slower in th social sciences, from 6,000 to 8,000 Ph.D.s per year (see Figures 3-1 and 3-2). Fueling much of this growth were rapidly rising levels of federal research funding, especially in the biomedical sciences through the National Institutes of Health (NIH). Departments and professors used these funds to support the graduate students and postdocs who provided labor in their labs.

But while "Ph.D. production has been zooming," Ginther continued, the number of tenure-track positions has grown only in engineering, where faculty ranks have increased by 50 percent over 30 years, As a result, Ginther surmised that almost half of new life science Ph.D.s and the "majority" in other fields have been finding work outside of academe. On campus, meanwhile, the "overwhelming majority" of new jobs in science has occurred in what Ginther called "non-track" positions that do not provide tenure opportunities. These include postdoc appointments as well as other kinds of soft-money research positions. Some of these jobs are essentially postdoc appointments by another, often ill-defined, title, such as staff scientist or research associate. Other positions, with names such as research assistant professor, do provide the opportunity to compete for grants as an independent investigator, but these positions are less secure because they are not eligible for tenure and rely primarily on external funding. Although these positions make it possible to

16 THE ARC OF THE ACADEMIC RESEARCH CAREER

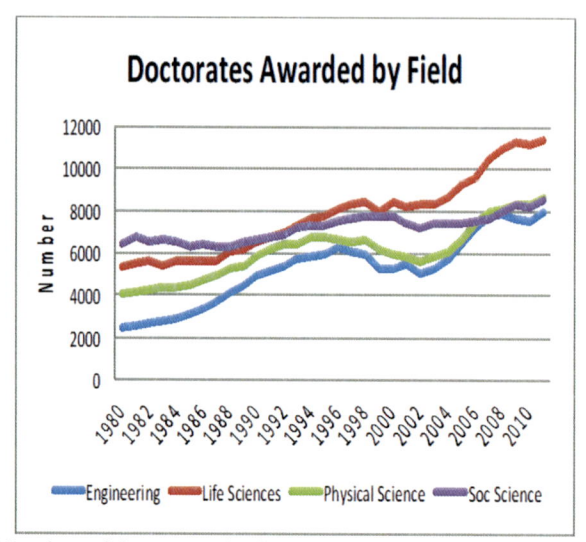

Figure 3-1 Number of doctorates in four broad fields (engineering, life sciences, physical sciences, and social science): 1980–2010.
SOURCE: 1980-2011 Survey of Earned Doctorates. National Science Foundation, Human Resources Statistics Program, National Center for Science and Engineering Statistics. Arlington, VA.

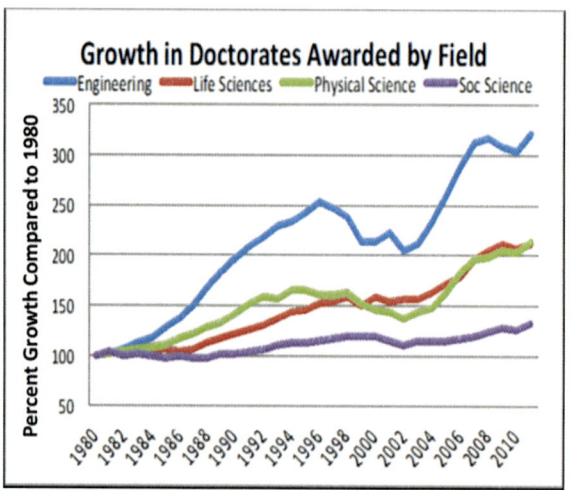

Figure 3-2 Growth rate of the number of doctorates in four broad fields (engineering, life sciences, physical sciences, and social science): 1980–2010.
SOURCE: 1980-2011 Survey of Earned Doctorates. National Science Foundation, Human Resources Statistics Program, National Center for Science and Engineering Statistics. Arlington, VA.

GETTING STARTED

participate in research, many participants believed that in general they provide less prestige, less independence, less job security, less income, and fewer benefits .

Fewer than 2,000 new biomedical Ph.D.s appear to have taken postdoc appointments in 1970, as opposed to nearly 6,000 in 2008 (see Figure 3-3). The current postdocs also stay in those positions longer. The percentage of new biomedical Ph.D.s taking postdocs "has hovered around 80 percent since the 1990s," said Ginther. As Ph.D. "numbers have increased,…you have this huge increase in the number of postdocs. You [also] see an increase in the number of behavioral and social [scientists] taking postdocs. Chemistry, [has had a] slight uptick, but not the huge increase in postdocs that you see in biomedical sciences."

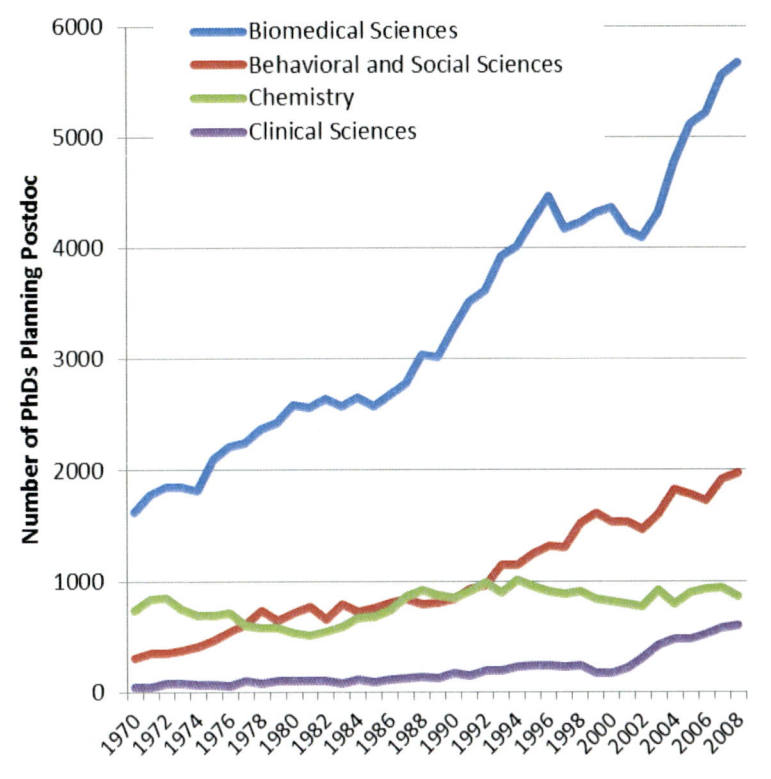

Figure 3-3 U.S. life sciences Ph.D.s with postdoctoral research plans compared to chemistry: 1970–2008.
SOURCE: Survey of Earned Doctorates, National Science Foundation, Human Resources Statistics Program, National Center for Science and Engineering Statistics. Arlington, VA; Prepared for the National Institutes of Health Advisory Committee to the Director Biomedical Workforce report (2012).

In a number of scientific fields, especially in life sciences, the prevalence of postdoc appointments means that for most students, earning a Ph.D. does not produce "a very high return" on the investment of time, effort, and opportunity involved in getting the degree, Ginther explained. Shortly after receiving the doctorate, "a Ph.D. in biological science makes 1.4 times what a BA in biological science or basic biomedical does"; this represents the lowest return of the four major science fields. This depressed income level, she said, is "driven by the postdoc," which can pay as little as $40,000 a year. Ph.D.s who leave academe at this point "have a higher return." "To take a postdoc," Ginther explained, "a student is making a large financial sacrifice." Another result of overproduction, Ginther added, is that "between 1997 and 2008, in almost every broad field, people are less likely to be using their Ph.D. to be employed in occupations that match their Ph.D. field. The one exception is math and computer sciences, and to a lesser extent…in psychology and social science."

In many fields, however, the postdoc has become entrenched as a de facto requirement for an assistant professorship, with Ph.D.s now routinely spending several years—including, according to Ginther, a "nontrivial percentage" who "have more than 8 years in a postdoc." Those who get to the tenure track today are therefore older than new assistant professors of past generations.

Moreover, even for the minority of scientists who do attain an assistant professorship, winning research funding has become more difficult than in past decades, in part because of the larger numbers of people, including many with non-tenure track university appointments such as research professorships, who are eligible to compete. "Since the 1990s, there has been a decoupling between the age of getting that first tenure-track job and the age of getting that first R01," the NIH grant that begins a scientist's career as an independent investigator, Ginther said (see Figure 3-4).

"My research has shown that the best way to get an NIH grant is to have already had an NIH grant," she continued. As a result, today, "the age of independence and an independent research career in biomedicine is over 40," as opposed to the situation several decades ago when scientists in their thirties and even younger often won their own funding. The rising age of "new" researchers became "so worrying that [Elias] Zerhouni, when he was head of NIH, instituted affirmative action for young investigators…Our young scholars were being shut out of the process in favor of older scholars" as funding became increasingly scarce and competitive, she said.

Overall, Ginther concluded, "It seems [that] in the life sciences, people are hanging out in postdocs, waiting for that academic job that…isn't going to happen…. We are taking the most productive years of our biomedical scientists' lives, and putting them in postdocs, where the opportunity to have an independent career is highly uncertain."

GETTING STARTED 19

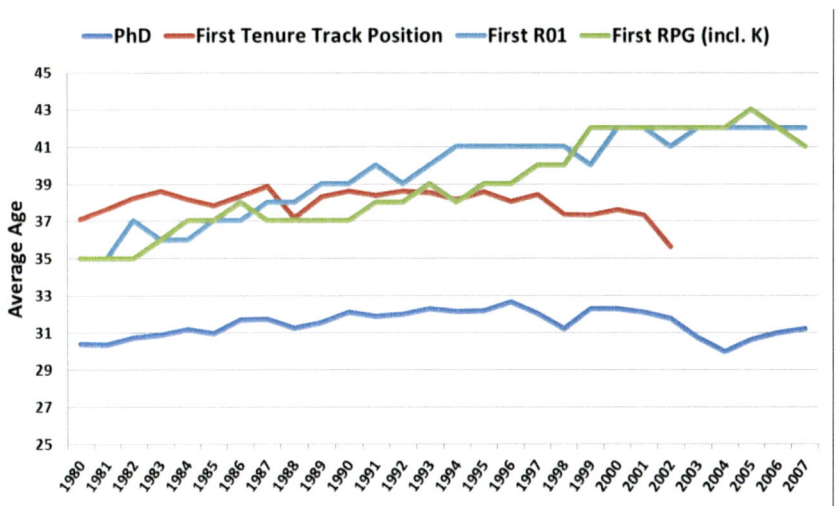

Figure 3-4 Average age at Ph.D., first tenure track position, and first R01/RPG (first major grant) for biomedical doctorates: 1980–2007.
SOURCE: Survey of Earned Doctorates, National Science Foundation, Human Resources Statistics Program, National Center for Science and Engineering Statistics. Arlington, VA; and National Institutes of Health IMPAC II; Prepared for the National Institutes of Health Advisory Committee to the Director Biomedical Workforce report (2012).

ASPIRATIONS AND REALITIES

Given these realities, what motivates new Ph.D.s to take postdoc appointments? How do they adapt to the fact that most do not emerge as assistant professors? "Do they know what they are getting into?" asked Henry Sauermann. "Do they actually know what these labor market odds are?" He and colleagues have investigated these questions in a number of surveys.

In looking at what he calls "the pre-beginning of the arc," he explained, his research examines the supply side of the academic labor market "not in terms of numbers of pieces, but in terms of people who actually have reasons to do these things. We want to understand what these reasons are because that also speaks to…what they are expecting and what society [has] to deliver to…fulfill its contract on the other side." Specifically, he wants to understand graduate students' and postdocs' "decisions to enter an academic career (or not)," their "preferences for various types of careers," and "how these preferences change over time." In addition, he has looked into postdocs' level of awareness of the job market, whether they regret " having done a postdoc," and "how [their] ex ante preferences and postcareer outcomes match up" over time.

"We heard earlier that a lot of people…enter nonacademic careers," he continued. But is that an "accident" and an undesired outcome, or "actually something they planned for as they start[ed] their Ph.D.s?" An initial survey sent to 30,000 postdocs and graduate students at 40 top-tier institutions and follow-up surveys of respondents 3 years later provided some answers.

"Implicit in many of our discussions is the assumption [that] people do Ph.D.s and postdocs because they want to be academics," Sauermann said. "A lot of advisors, from what we hear, have that assumption" and often "imposed" it on their students. But, he emphasized, expanding on Ginther's opening point, in examining motivations as well as labor markets, "we really have to think across different fields. [The preferred career] is not the same in the life sciences as it is in engineering," where many people choose to work in industry. "We know the actual career paths are different, and it is not surprising that the preferences are quite different, as well."

Asked to pick their preferred career from choices that included a faculty appointment that emphasizes teaching, a faculty appointment that emphasizes research, government work, a job in an established private firm, a job in a start-up company, or some other career, Ph.D.s in life sciences, chemistry, physics, and engineering expressed different preferences. By wide margins the life scientists and physicists chose faculty with a research emphasis first and faculty with a teaching emphasis as their second choice. The chemists and engineers, however, named work in an established firm as their top preference, although each group also chose faculty as their second choice, the chemists preferring a teaching emphasis and the engineers choosing a research emphasis. "The point is," Sauermann says, that these Ph.D.s have "a very broad range of career interests. Academia is an important career interest for these people, [but] it is not the only one."

Scientists' career interests are not only broad, Sauermann's research shows, but also change over time. In 2010 and again in 2013, the same group of graduate students in life sciences, chemistry, physics, and engineering rated the attractiveness of careers as university faculty with an emphasis on research, in government, in an established firm, or elsewhere. Regardless of discipline, the allure of the faculty post declined during the interval, with the steepest decline among life scientists and physicists. The desirability of working for an established firm rose for the life scientist and chemists and remained unchanged for the physicists. In all groups, "other," which might include work for a start-up firm or other endeavors altogether, also rose.

What dulled the lure of the academic life? Some insight comes from open-ended comments by respondents who had initially named a research-oriented professorship as their first choice, Sauermann said. He offered a few "nonrandom" examples of the most pointed comments:

- "Realizing that university faculty usually spend most of their time on activities supporting the research but not the research itself,"
- "I got tired of doing work that only matters to a handful of people, has no impact on society, and pays poorly,"

- "I don't want to be my advisor. Ever."
- "I've discovered that I'm a mediocre scientist but a really good teacher, and teaching makes me happy."
- "The realization that very few tenure-track positions actually exist, the lack of research funding, and too many qualified applicants."

A common theme among the graduate students' comments, Sauermann noted, was their realization that "academic research is not actually doing academic research [but] managing, fundraising, struggling with grants, and all these issues.... A lot of these people go into science because they want to be at the bench, study stuff, discover stuff and not just tell other people what to study.... For some of these students, that seems to be one of the insights that they learn while they are watching their advisors and seeing how it plays out." A number of people also wrote, "I want to start my own company, or I want to go work for a big pharma company, because that is where I know stuff actually happens and goes down the pipe[line], and hits people in terms of...[new] drugs..." Someone else, however, said, "I can make the biggest impact providing a really cool, important paper." In Sauermann's opinion, "both are...important impacts, but we have to understand that people see these things in different ways," depending on their own values and experiences.

The career preferences that postdocs expressed were a bit different from those of the graduate students, however, perhaps because people with nonacademic bents had already taken other directions after finishing their Ph.D. Asking life sciences, chemistry, physics, and engineering postdocs to choose their most preferred career from among a faculty job emphasizing teaching, a faculty job emphasizing research, and working for government, an established firm, a start-up firm, or other, showed that faculty with a research emphasis led in all groups by very wide margins.

Given the very tight academic job market, however, do Ph.D.s enter their postdoc appointments aware that they face such low odds of ending up in the faculty job they hope for? Perhaps, Sauermann suggested, "they don't know what the labor market conditions are. Maybe they know, but they think they are at the top end of the distribution. Maybe they actually do a postdoc for other reasons, [not intending] to become the academic researcher.... Maybe they actually want to do something else."

Asked what percentage of Ph.D.s in their field land a tenure-track job within 5 years of earning the doctorate, the four groups of postdocs surveyed each gave estimates that on average came within 3 percentage points of the actual figures, which in these four broad categories stand only in the mid-teens. "They are smart people, and they read *Science* Careers," Sauermann said. "They read the discussions; they know quite well that the odds of getting a position are not that great."

But does this mean they've given up hope of getting one of those coveted posts for themselves? The majority of the postdocs—61 percent of the life scientists, 55 percent of the chemists, 63 percent of the physicists, and 47 percent of the engineers— said that having an academic research career was indeed their motive for pursuing a postdoc. Despite their accurate knowledge of the job market, those aspiring to a tenure-track job still viewed their own chances of having one within 5 years as quite a bit better than the low averages they had quoted; they estimated a 55 percent chance for the life scientists, 68 percent for the chemists, 42 percent for the physicists, and 58 percent for the engineers.

"The labor market expectations, at the level of the global labor market, seem to be quite well calibrated," Sauermann noted, but "maybe not so much at the level of the individual person....Any given person could truly be the next Nobel Laureate, right? We can't tell them 'No, you are overestimating your chances.' Any given person might be right, but we know that, collectively, they can't all be right."

Advisors, however, often have considerable influence over the aspiring scientists they teach. "We found that about 80 percent of students [say] that my advisor most strongly supports an academic research career, even though we know most of them don't get those careers." Young scientists, Sauermann continued, need "a broader set of information sources…to understand labor market and the different kinds of careers [available to them] and make informed career decisions."

But society also needs to know which students end up pursuing which kinds of careers, Sauermann added. "Which students get the jobs they want? Which ones don't?...How is it related to ability? Is it that the really, really smart people in the end get the academic positions? Is it the ones with the lowest opportunity costs, who have nothing to lose as they just stick around in the postdocs? Is it people with the biggest persistence, who are just determined to do it, and they stick around and don't get lured away by something else?"

Beyond that, "How selective do we want the [career selection] mechanism to be? What happens to people who don't get selected or don't select into these different careers? How can we evaluate the career outcomes we see?"

FAMILY MATTERS

Despite the complex motives that Sauermann's research revealed, the reason that a great many postdocs lose interest in careers as research-oriented faculty members is simple and straightforward, Mason found in her surveys of the University of California-Berkeley faculty. Her recent book is entitled *Do Babies Matter?: Gender and Family in the Ivory Tower*,[6] and the answer, at least for women in the early stages of academic careers—meaning before tenure—is that the desire to start a family must often compete with the desire to have an academic research career. As Girgus also noted, "graduate students and

---

[6] Mary Ann Mason, Nicholas H. Wolfinger, and Marc Goulden. *Do Babies Matter?: Gender and Family in the Ivory Tower*. Rutgers University Press (2013).

postdocs, like faculty, are often trying to integrate a productive work life with family formation, something that we have come to recognize as important only recently."

The conflicting desires are most apparent in the life sciences, where women now earn more than half the Ph.D.s. And because of overproduction of Ph.D.s fueled by NIH funding and the use of graduate students as low-cost lab workers, life sciences is also the "hardest in terms of getting a [faculty] job" and where young scientists face the greatest risk of "staying in your postdoc forever." Life sciences is therefore the field "most hard hit by all the demographics that I am going to be telling you [about]," but the same trends exist in the other fields as well, she said.

Moving from the postdoc to the tenure track is unlikely for the great majority of scientists, but is unlikeliest of all for married mothers, who are 35 percent less likely than married fathers to make that jump, Mason said. "The major slide of women out of the academic pipeline after getting their Ph.D.s occurs before they get their tenure-track job, usually in the postdoc years, or they have changed their minds in the graduate student years," Mason continued.

When postdocs become parents, "twice as many women [as] men are likely to change their career goal away from being a research professor… . Forty-one percent of women who had babies when they are postdocs say that they no longer want to have a research career, as opposed to 20 percent of men." The trend is even sharper, and the difference from men is even greater, among women who have babies while in graduate school, she added. Many "babies [are] born during the postdoc years," which, given postdocs' ages—generally the early to middle thirties—"is natural," especially since fewer women have babies while graduate students, at least at research universities, she said. When asked why they changed their career plans, female graduate students at the University of California-Berkeley "almost uniformly said for family reasons. 'I want to be able to enjoy having a family, [being a] mother and wife, which are close to impossible when one chooses academia. The clock is ticking, and it doesn't stop for anything or anyone.' "

"The main problem for all women in academia who have babies is they just don't have enough time," Mason said. Berkeley postdocs who are mothers average 100 hours of work a week: 43.6 doing research, 3 teaching, 15.9 doing housework, and 36.3—almost a second full-time job—giving care. Postdoc fathers put in more time in the lab—51.6 hours—but only 2.4 on teaching. And they put in strikingly fewer hours working at home—12.6 on housework and 20.7 in child care. Regardless of gender, postdocs without children, on the other hand, work a "mere" 80 hours a week, more than 50 of them in the lab (see Figure 3-5).

Beyond a crushing workload that includes less time spent in professional competition, postdoc mothers often face another major disadvantage. At the many universities that classify them as trainees rather than employees, "they are not covered by the [Family and Medical Leave Act, which provides unpaid time

Everybody is Busy (UC Postdoctoral Scholars)

	Women with Children (N=211)	Men with Children (N=394)	Women without Children (N=704)	Men without Children (N=884)
Caregiving	36.3 hrs	20.7 hrs	6.7 hrs	7.5 hrs
Housework	15.9 hrs	12.6 hrs	11.5 hrs	10.8 hrs
Teaching	3 hrs	2.4 hrs	3.2 hrs	2.5 hrs
Research activities	43.6 hrs	51.6 hrs	51.6 hrs	54.2 hrs

Source: Marc Goulden, Mary Ann Mason, and Karie Frasch. 2009. "UC Postdoctoral Career Life Survey." (http://ucfamilyedge.berkeley.edu/grad%20life%20survey.html).

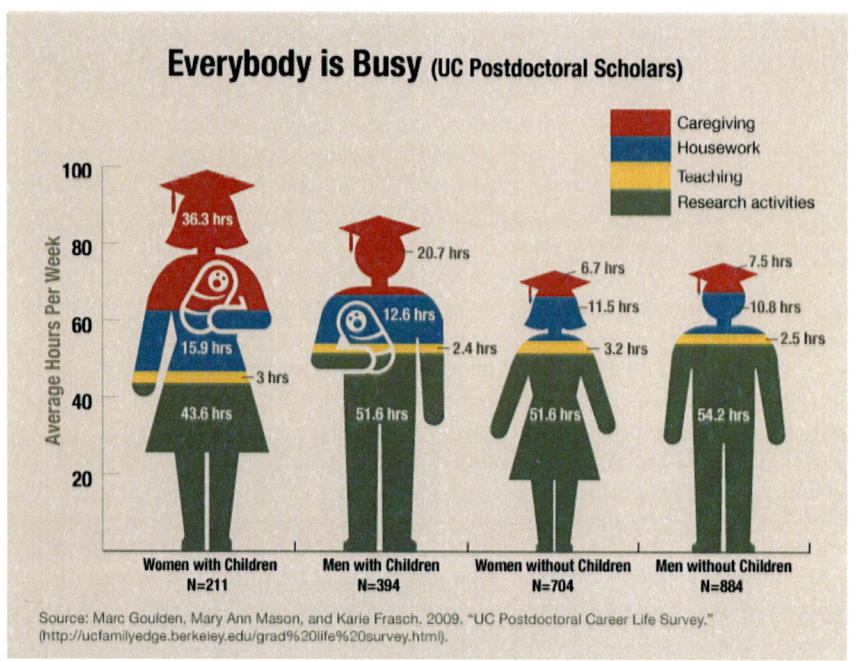

Figure 3-5 Workload by type of work (caregiving, housework, teaching, and research activities) for U.S. postdoctoral scholars with and without children, by gender: 2009.
SOURCE: Goulden, M., M. A. Mason, and K. Frasch. "UC Postdoctoral Career Life Survey". Berkeley, CA: UC Berkeley, (2009). Available at http://ucfamilyedge.berkeley.edu/grad%20life%20survey.html.

off to care for family members, or by] Title 7," which bans employment discrimination based on pregnancy. Some universities, however, include graduate students and postdocs in the maternity leave plans that they provide to faculty members. The University of California, for example, is among the minority of universities that give graduate students and postdocs at least 6 weeks of paid maternity leave (see Figure 3-6). Princeton provides them a range of benefits, Girgus noted, including assistance paying for child care, backup child care in case of illness, financial assistance to provide child care during professional travel, and more. Most universities, however, can afford to provide only a good deal less. Said Mason, "There is a lot of room for improvement at the graduate student and postdoc years, if we don't want to lose women out of the pipeline."

Provision of PAID MATERNITY LEAVE for Academic Poplations at Association of American Univeristies (AAU)

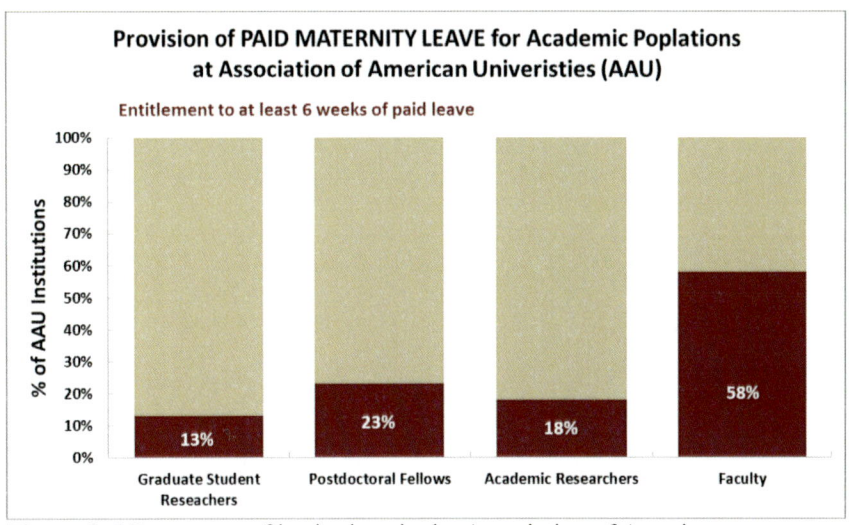

Entitlement to at least 6 weeks of paid leave

- Graduate Student Reseachers: 13%
- Postdoctoral Fellows: 23%
- Academic Researchers: 18%
- Faculty: 58%

Figure 3-6 Percentage of institutions in the Association of American Universities (N = 62) which provide at least 6 weeks of paid maternity leave for academic populations: 2008.
SOURCE: Mason, M. A., M. Goulden, and K. Frasch. "Family Accommodation Policies for Researchers at AAU Universities Survey." (2008).

4
The Tenure Track and Beyond

Many of the issues that aspiring academic scientists encounter in graduate school and postdoc days persist and even sharpen as they reach the tenure track. As was noted by many of the participants, some of the former graduate students and postdocs who do not succeed in making this jump of course seek careers in nonacademic fields. Substantial numbers remain in academe, however, as "non-track" or "off-track," adjunct, contingent, or part-time faculty who are hired on a temporary basis to give either individual courses or full teaching loads. Consisting "disproportionately [of] women Ph.D.s with children," this "cheap labor force" provides an increasing share of undergraduate instruction nationwide, Mason said. The ranks of the part-time faculty, Valerie Martin Conley, professor of counseling and higher education and co-director of the Center for Higher Education at Ohio University, noted, have grown much faster than those of full-time faculty (see Figure 4-1).

In "2009, we had gotten to a point where we had a one-to-one ratio of [full- to part-time] people who were employed as faculty, [doing] instruction, research, or service, at all postsecondary institutions in the U.S.," Martin Conley said. "Most research data…[show] that there are now actually more people employed part-time than full-time in the category 'faculty,' " she added. Many part-time instructors entirely lack health, retirement, and other benefits from the universities where they work. Full-time contingent faculty members may receive some or all of those benefits, but often less generous plans than tenured and tenure-track colleagues enjoy.

Those faculty members who have taken the crucial step that offers them the chance of attaining the security, permanence, and status provided by tenure now turn their focus toward winning that goal, an effort that can, depending on the institution, last 7 or even 10 years. Success demands great amounts of time and energy to amass a record of publications and successful grant applications, as

"Adjunctification"

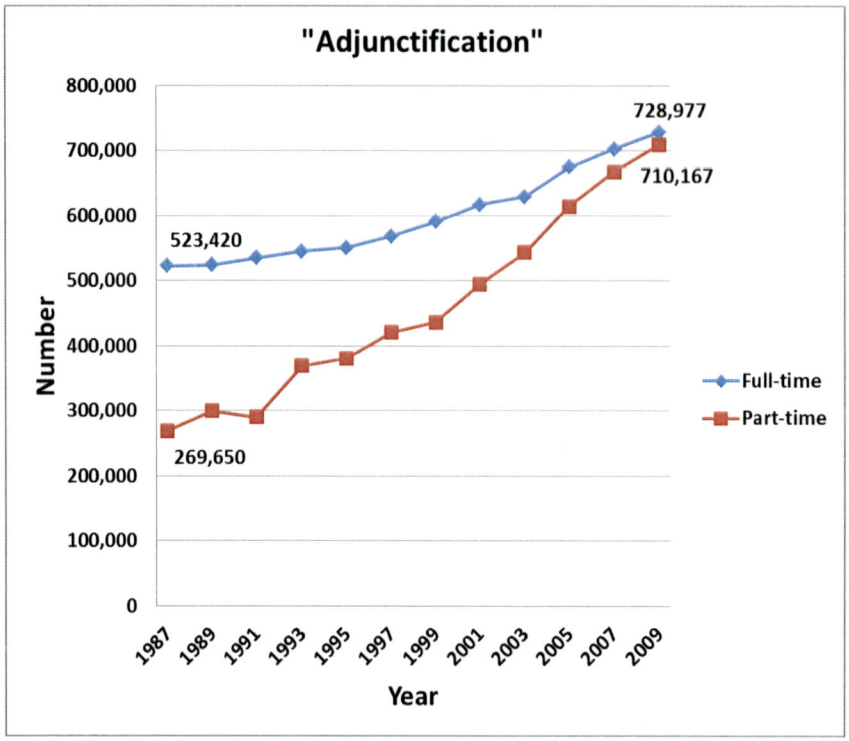

Figure 4-1 Number of full-time (tenured, tenure-track, and non-track positions) and part-time faculty: 1987–2009.
NOTE: For further detail, see figure 2-3.
SOURCE: U.S. Department of Education, National Center for Education Statistics, Higher Education General Information Survey (HEGIS), Employees in Institutions of Higher Education, 1970 and 1972, and "Staff Survey" 1976; Projections of Education Statistics to 2000; Integrated Postsecondary Education Data System (IPEDS), "Fall Staff Survey" (IPEDS-S:87-99); IPEDS Winter 2001-02 through Winter 2011-12, Human Resources component, Fall Staff section; and U.S. Equal Employment Opportunity Commission, Higher Education Staff Information Survey (EEO-6), 1977, 1981, and 1983., Table 290, (This table was prepared July 2012).

well as effective teaching and institutional service, to meet their particular department's and university's standard of productivity.

Because assistant professors are often in their mid-thirties or older, however, most are also coping with the conflicts and contradictions of pursuing demanding careers while fulfilling family responsibilities, often in a household with two working spouses or partners. "We discovered that more than 70 percent of both tenured and tenure-track faculty were partnered and more than 75 percent of those had working partners," Joan Girgus said. Unlike the faculty members of Shirley Malcom's graduate school days, very few today can rely on a spouse at home to manage household and family matters. "Furthermore," Girgus continued, "more than half our faculty—male and female—reported having ongoing care responsibility for children under the age [of] 18."

As highly educated professionals, academics and aspiring academics also tend to marry other highly educated people; female scientists show a particular propensity to marry other scientists. This adds the extra complication that the spouses of many academics also have serious careers of their own, whether on campus or off, Girgus noted. Making the move to new universities, which is often needed to advance an academic career, also often requires moving to a new city or region. Accommodating the spouse's career can thus pose a difficult and even insuperable challenge, both for universities seeking to hire new faculty and for faculty couples wishing to stay together while pursuing both professional interests. These days, two-city academic marriages are far from rare.

"The dual-career issue is a major problem," said Carol Hoffman, associate provost and director of the Work/Life program at Columbia University, which happens to be in the nation's largest metropolis. "It is easier in a city like New York than it is in some rural smaller communities." In more bucolic Princeton, on the other hand, "dual-career issues have been and continue to be the most challenging, both for faculty members and for the university," Girgus said. Finding acceptable solutions is "crucial for faculty being recruited from other universities." Indeed, suggests Mary Ann Mason, the complications of pursuing an academic career while part of an academic couple may explain the "odd finding" that "single mothers actually do better than married mothers in terms of getting tenure," because they are freer than women with spouses to make the moves that are best for their own careers.

In past generations, university nepotism rules often forbade spouses from working in the same department or even at the same institution, and women's aspirations often took the back seat to their husbands' careers, Malcom observed. More recently, however, as Girgus and Hoffman noted, universities' attitudes have changed to match the reality of today's many two-career households, and now institutions try to help find suitable positions for the spouses of faculty relocating to their campuses, either in their own or neighboring academic institutions or with other employers in the surrounding community.

IT'S ALL IN THE TIMING

As female assistant professors undertake the drive for tenure, however, a factor even more important than the quality and quantity of their research can determine their ultimate success: whether or not they have a baby soon after earning their Ph.D.. Between 1979 and 1995, only 53 percent of women on the tenure track with "early" offspring—defined as those born within 5 years of the mother's Ph.D.—won tenure, said Mason, citing information from the Survey of

Doctoral Recipients. Men with early babies, on the other hand, attained tenure 77 percent of the time, and women with "late" babies—a first child born 5 or more years after the Ph.D.—did so 65 percent of the time. Also, married mothers were 35 percent less likely than married fathers to get onto the tenure track in the first place.

The pattern of long work hours—and extra-long hours for mothers—of course follows scientists onto the tenure track. Among faculty members at the University of California-Berkeley aged between 30 and 50, Mason said, mothers put in the longest work week—just over 100 hours—but spend fewer hours on professional tasks—51.2—than any category of colleague. Faculty fathers come second in working time—just under 90 hours a week in total—but spend 4.2 more hours on professional work than mothers. They also spend 15.2 fewer hours a week on caregiving and 2.7 fewer on housework.

Non-parents of both genders meanwhile keep quite similar, and less stringent, schedules: nearly 80 hours of total work a week, 60 of them devoted to professional work, about 10 to housework, and about 8 to caregiving. "Some things are changing, but…some things aren't changing very much," Mason said. "The women are still bearing the brunt of the housework for sure." As these patterns show, "Research tends to be the piece that you can, in fact, shorten if you run out of time because of family life," Girgus observed. "That can have serious consequences" for one's career.

Given the terrible press of time on academic mothers, "One of our science department's chairs said to me [that] he tells women to wait until tenure before they start having a family," said Cathy Trower, research director of the Collaborative on Academic Careers in Higher Education at the Harvard School of Education. But, she added, "if [women] wait to have a baby until they get tenure, if it is 9 years or 10 years on the tenure track, it probably will be too late."

The intense pressure to achieve tenure is also a probable reason that tenure-track women have fewer children than comparable men. In 2003, for example, Mason said, 73 percent of female assistant professors at the University of California-Berkeley were childless, as opposed to 61 percent of the males. Similar percentages of women and men had a single child—15 percent and 17 percent, respectively—but men were about twice as likely to have larger families—12 percent with two children and 10 percent with three or more, as opposed to 7 percent of women with two children and 5 percent with three or more.

Recognizing that the quest for tenure and the desire to have children are sometimes in conflict and wanting to increase the diversity of their faculties by hiring and retaining qualified women, many universities have adopted programs to help faculty members balance their career and family responsibilities. These often include the ability to stop the tenure clock for a set period, assistance with child care, and other services (see Box 4-1). At Princeton, which has an extensive range of professional and family policies to aid assistant professors, and which endeavors to "do all that it can to help assistant professors get tenure," Girgus said, "men and women assistant professors at Princeton receive tenure at the same rate and have for at least 30 years." But, as John Tully, Sterling Professor of Chemistry and a professor of physics and applied physics at Yale University, noted, "Many of the possibilities that we have at elite institutions might not be true for most of the academic faculty in academic careers."

Box 4-1
Examples of Family Support Policies

Princeton University
- Maternity leave
- Automatic 1-year extension of the tenure clock for each child
- Work-load relief for the primary caretaker
- Backup care program
- Dependent care travel fund
- Employee Child Care Assistance Program
- Expanding on-campus child care
- Employee Assistance Provider Work/Life Program
- Partner placement assistance
- Tuition grants for college-aged children

Columbia University
- Part-time Career Appointment
- Maternity Disability
- Child Care Leave
- Parental Workload Relief
- Tenure Clock Stoppage (for parental reasons)
- Assistance finding housing, schools, and other needed services, including partner placement

---

Family-friendly policies also appear to have taken some of the pressure off young faculty at the University of California-Berkeley as well and to have encouraged more women to have children before reaching tenure, Mason suggested. In 2009, the birth of new babies was up two-thirds over 2003 for female assistant professors and up 20 percent for males. Faculty fathers, however, still had larger families.

More detailed information about the effects and utility of family policies, Hoffman told the workshop, comes from a study by Columbia University aimed at seeing both whether its family policies have accomplished their intended purpose and whether complaints that some people abuse the policies are valid. Looking at records from 1990 to 2008 for tenured and tenure-track faculty from across the university (except for the medical center), the study took note of individuals' department, gender, age, dates when they made use of a policy, how many times they used the policies, and how their careers progressed. The study identified 167 faculty members who had availed themselves of one or more parental policies during those years, 42 of whom were tenured and the rest on the tenure track. Significantly, only 14 percent of users came from natural science departments and 4 percent from engineering, perhaps indicating the low numbers of women in those fields, Hoffman said. Social science departments accounted for 38 percent of the users; professional schools, 23 percent; arts and humanities, 21 percent.

The policy most popular among both tenure-track and tenured faculty is parental workload release, which removes teaching responsibilities for a set period. "Tenure-track women use the workload relief in similar numbers to men," Hoffman said, noting that this represents "a way greater percentage of women," because they constituted only 30 percent of the faculty overall. Among people with tenure, however, 31 men and 9 women used teaching relief, numbers "more proportionate" to the mix of genders in the overall faculty.

Men and women both made their first use of teaching relief on average in the middle of their fourth year after being hired. The 51 women who used it while on the tenure track at Columbia "ranged from 29.4 years old to 42.4 with an average of 36.6 for the tenure track, a pretty high age" for a first child, Hoffman said. The nine women past tenure making their first use were an average of 42 years old, and Hoffman observed that "the chance of maintaining fertility at that age is actually quite slim."

Thirty-two faculty members, 20 of them before tenure, made use of two periods of teaching relief. The numbers for the two genders were "almost the same," with men forming a small majority. Hoffman believes that this provides further corroboration for "work at Berkeley and other places [showing] that men are having more children than women during their academic career." More evidence, Hoffman added, appears to come from the 69 faculty parents adding new children to Columbia's health plan during 2007 and 2008, 54 of whom were men and 15 of whom were women. Only 2 of the mothers had attained tenure and 13 were still on the tenure track. These figures do not provide definite proof, however, Hoffman noted, because some of the parents may have availed themselves of health plans provided by a spouse's job.

Did teaching relief affect tenure outcomes? The study had no control group, so no conclusion was possible, Hoffman noted. Despite this, "we were pleased to see that a lot of folks, in fact, did proceed with tenure." When the study closed in 2008, however, 49 percent of the users were still on tenure track at Columbia or other universities, and another 43 percent have attained tenure, 25 percent at Columbia. But the study also corroborated another conclusion of "the research from Berkeley, that more women than men left their tenure track and tenured positions...and went into research positions,...administrative positions,...adjunct faculty part-time, faculty lecturer positions, but no longer were [on] tenure track or tenured."

Over time, the study revealed, the use of teaching relief has become "normalized" at Columbia; in 2008 alone, 28 individuals used it. Though the earliest users of family policies were "a handful of brave women in the seventies and eighties, and early nineties," Hoffman said, it is now "just sort of understood that if you have a child, you use teaching relief."

But even if family policies help women achieve tenure, the effort "must be [taking] a greater toll on their own selves and well-being because they have [fewer] hours for themselves overall," because of the heavy load of child care and household responsibilities that so many bear. "It doesn't look like women aren't getting tenure, but," Hoffman asked, "what is it that it is taking out of them?"

"In the United States,...we don't have [national] maternity or parental leave policies, and we don't have affordable early child care starting at the earliest stage," Hoffman continued. "Most developed countries have either lengthy initial year or two of time off when you have a child, or have free and early child care. For scientists, full-time long-day full-year child care is absolutely imperative, because your labs run like that. You don't get summers off...[Child care] is a national issue. Universities can't solve it alone."

The demands involved in caring for children may, in fact, account for what Girgus called "one of the very few differences between men and women faculty that we have at Princeton"—an extremely wealthy institution with superior career and family supports for faculty—"women spend longer as associate professors than men do" before winning advancement to full professor. One possible reason, she suggested, may be that "they are more reluctant to stand for promotion." But another plausible explanation could be that "a lot of women faculty, a fair percentage, do wait until they have tenure before they have children," she said.

AFTER TENURE

Apart from that, though, associate professor status generally constitutes "a continuation," Girgus said. Professors continue with their research and other duties. And it is now clear that an individual will likely be at the institution until retirement, unless lured away by a more attractive offer from elsewhere. For this reason, "we believe it is crucial that faculty begin seriously planning for retirement at this point," although, she said, many decline to do so.

At the same time, promotion to associate professor is also the point when invitations to move to other institutions begin to arrive in earnest. "Once they get tenure, that is when they get heavily recruited. We have a big spike up in outside offers around age 40, 45," said Marc Goulden, the director of data initiatives in the Office for Faculty Equity & Welfare, University of California-Berkeley, who conducted a study of the University of California's Voluntary Early Retirement Incentive Program. When this happens, Girgus notes, "dual-career issues get, if anything, more challenging."

For those who remain at their original institution, however, "a rule of thumb is roughly 6 years on the tenure track, and then another 6 at associate before standing for full professorship," Trower says. "However, research shows that many faculty remain at the rank of associate either because they never apply for full, or because they do and they are denied, and they stay."

Research also shows that the longer people remain in associate status, the less satisfaction they feel about their working conditions and workplace, according to Trower, who reported on a study from the Collaborative on Academic Careers in Higher Education (COACHE) of 1,263 tenured associate professors in the science, technology, engineering, and mathematics fields at a range of universities.[7] The findings indicated that "people are pretty happy right after they get tenure," she said. "Then, as they go through at associate, …at 7 to 12 years in"—at the point, in other words, where promotion to full professor may be expected—"there is a lot less satisfaction with the workplace." Across the arc of a career, she said, "faculty job satisfaction follows a U. It starts high and it drops, drops, drops, drops in this trough. Then, it picks up again toward the end of one's career."

During the associate professorship years, she continued, faculty members often also begin to receive requests to take on serious service and leadership roles, such as chairing an important committee or the department. Such duties, while significant to the institution, take time from research and do not add to the person's scientific reputation or grant support. Women in particular often report feeling that they must bear an inequitable share of the load and sense a "certain cultural taxation on women in the academy to do committee [or other] service," she said. Research shows that men "will say no to leading the department until they get to full. Many women, [on the other hand], are tapped to be a director or a department chair."

A "counterintuitive but interesting finding," given all the research about the stress of academic motherhood, is that childless women associate professors express less satisfaction than those who are mothers. Possibly, Trower suggested, there might "actually be a pendulum shift, with all the focus now on people with kids," leaving those who are not parents feeling overlooked. Or perhaps, having given up family for career, they feel disappointment in the outcome.

Then, finally, those who reach the top rank and find themselves "professor at last," enter the "time to give back," Girgus said. At Princeton, "we really

---

[7] More information can be found on the COACHE homepage, available at: http://isites.harvard.edu/icb/icb.do?keyword=coache&pageid=icb.page307142

begin to jump on our full professors…. The minute you become a full professor, you are departmental representative or you are the director of graduate studies or you are chair of the department. There is a long list because we have a lot of faculty governance and a small faculty." Beyond that, for academics who have attained eminence in their research fields, invitations to move continue to arrive. "We find dual-career issues here are absolutely crucial, both for recruitment and retention. Bringing people in from other universities or persuading them to stay at Princeton, it is the central piece of what we do in recruitment and retention, the central difficult piece, I should say."

The issues of the later career underline another striking trend in higher education, what Martin Conley called the overall "graying" of the faculty. "Both full-time and part-time faculty are aging, as is the population," she said. Between 1988 and 2004, "the average age of part-time faculty has increased from 44 to 50 years" and of full-time faculty from 47 to 50 years, according to the National Study of Postsecondary Faculty, she said (see Figure 4-2). In 1987, the percentage of full-time instructional faculty under the age of 40 slightly exceeded that of full-time faculty over 55, at 25.2 and 24.2 percent, respectively. By 2003, the older cohort was almost twice the size of the younger, at 34.9 percent as opposed to 19.2, respectively.

One reason may be that people "are living longer, healthier lives" and probably fewer are dying young. But another is that, starting in the 1990s,

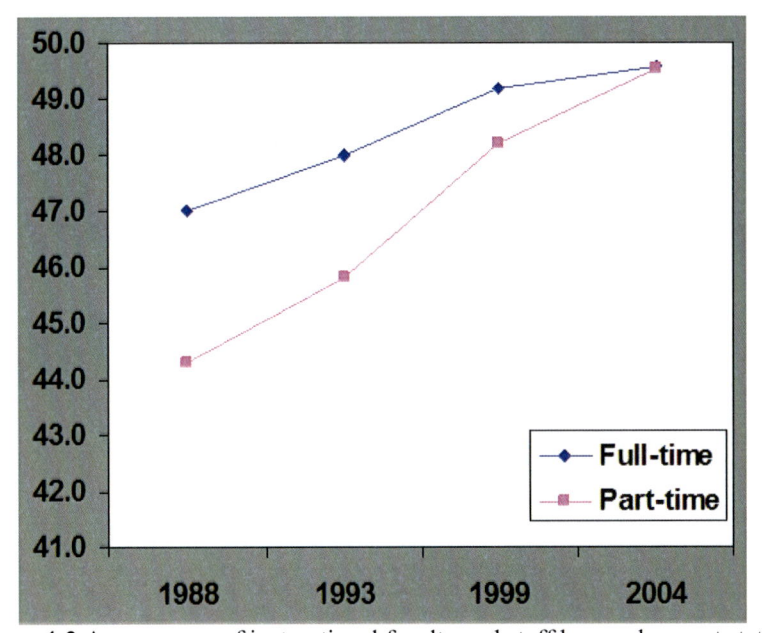

Figure 4-2 Average age of instructional faculty and staff by employment status at 4-year institutions for selected years: 1988, 1993, 1999, and 2004.

SOURCE: U.S. Department of Education. Institute of Education Sciences, National Center for Education Statistics, *National Study of Postsecondary Faculty (NSOPF)*.

faculty have been working longer. Between fall 1992 and 1998—with the end of mandatory retirement coming in 1994— the percentage of departures by full-time faculty because of retirement declined, giving "one of our first clues that faculty were beginning to delay retirement," Martin Conley said.

The age structure of the faculty at UC Berkeley, for example, shows an unmistakable trend toward growth among the oldest cohorts, according to data that Goulden supplied (see Figure 4-3). In 1979, faculty members under 34 years of age numbered 140 and those over 60, only 173. In 2013, the 97 faculty members over 70 outnumbered the 92 between 30 and 34, with only 11 aged below 30, as opposed to 52 in that youngest age bracket in 1979. Today, the 342 faculty members between 60 and 69 substantially outnumber the 253 between 30 and 39.

As faculty move into the later years of their careers, "retirement planning…moves to the forefront" in preparation for what Janette Brown, executive director of both the Emeriti Center at the University of Southern California (USC) and the Association of Retirement Organizations in Higher Education (AROHE), calls "the new life stage, which is roughly between the ages of 60 and 85."

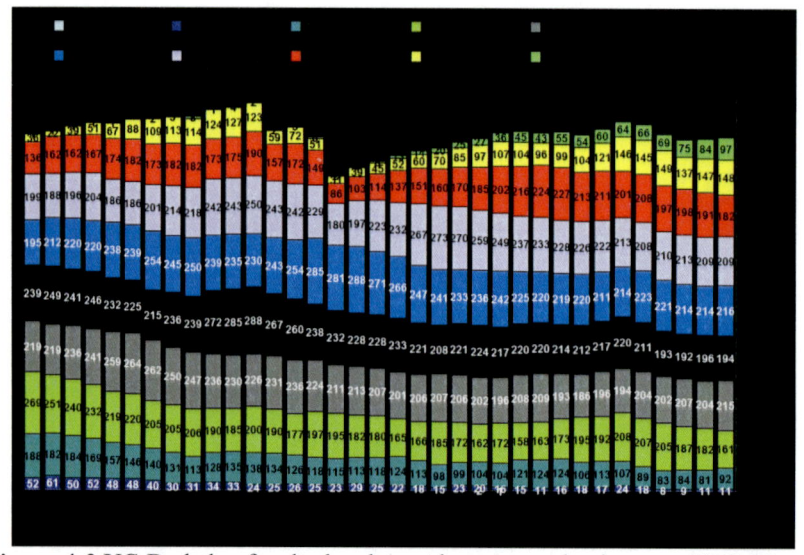

Figure 4-3 UC-Berkeley faculty headcount by age: academic years 1979-80—2013-14.
NOTE: Data for academic year 2013-14 is preliminary.
SOURCE: UCB Faculty Personnel Records, AY 1979-80—2013-14. Prepared by Goulden, September 2013; updated October 2013.

5
Moving into Retirement

Until January 1, 1994, the day that the amended Age Discrimination in Employment Act forbade higher education institutions from imposing a mandatory retirement age on faculty, academic careers almost always ended at or before the age of 70. Only a handful of faculty members continued working longer, generally through special arrangements that exempted them from the mandatory retirement rule. Under the amended law, tenured professors' lifetime appointments thenceforth permitted them to work as long as they wished, making when they retired into a matter of personal choice rather than of university policy.

The change created new challenges for both institutions and individuals. Universities now had to navigate the potentially sensitive process of separating long-serving faculty members from the academic careers they had pursued for a lifetime. Professors now had to decide when to stop doing the work that shaped their lives and, for many, Edie Goldenberg noted, defined their identities.

Twenty years after the new law, these issues have increasing salience on campuses across the country, because the cohorts that are nearing or have reached beyond the traditional retirement age represent an increasing proportion of the academic population. Between fall 1992 and 1998, the percentage of departures by full-time faculty because of retirement declined, at "almost every type of institution," said Valerie Martin Conley. Continuing to work into their seventies became more commonplace. At the University of California-Berkeley, for example, "between 2002-3 and 2013, several faculty members have separated from the university as late as age 83," Marc Goulden said.

Universities are now seeing a "wide diversity of retirement and retirees," a trend that will only grow, Martin Conley said. According to one study, the average age at which faculty members expect to retire is 66.5; for those still working at or past 71, however, the expected age is 75.7. Those figures, however, hide a great deal of variation. According to the 2003-4 National Study of Postsecondary Faculty survey (see Figure 5-1), a plurality—37.3 percent—plan to retire "on time," presumably in their mid- to late sixties. The next largest group, 28.6 percent, says they will retire "late," and a slightly smaller share, 25 percent, intends to do so "early." Almost 8 percent, however, say they will work until "very late," and a tiny sliver, 1.3 percent, intends to leave "very early."

More recently, a 2011 study of full-time faculty members over age 60 found fully three-quarters expecting to continue working past the customary retirement age, with 60 percent saying they would do so by choice and 15 percent citing external factors, which were "primarily financial" according to Martin Conley.[8] Because these data were collected shortly after the 2008 economic collapse, however, many people's retirement accounts may have since rebounded, she added.

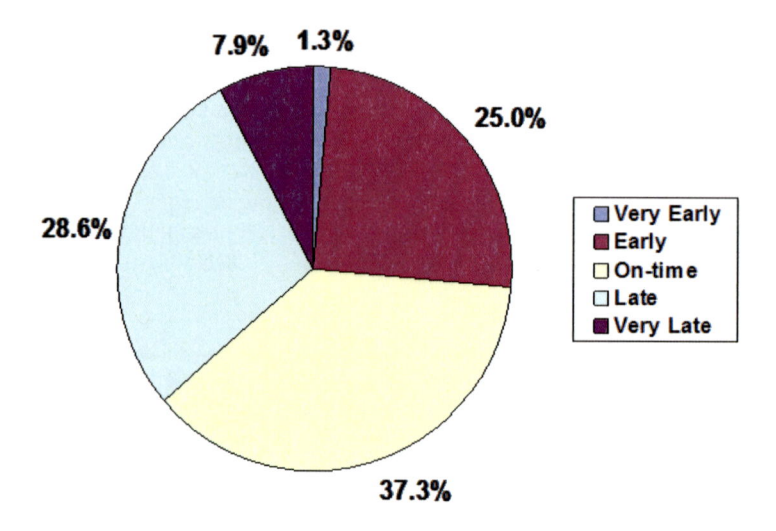

Figure 5-1 Percentage distribution of expected timing of retirement of full-time instructional faculty and staff: 2003–04.
SOURCE: U.S. Department of Education. Institute of Education Sciences, National Center for Education Statistics, *National Study of Postsecondary Faculty (NSOPF), Data Analysis System (DAS)*.

---

[8] Paul J. Yakoboski. "Should I Stay or Should I Go? The Faculty Retirement Decision." *Trends and Issues*, TIAA-CREF Institute (December 2011).

"Almost all eligible full-time faculty members participate in employer-sponsored retirement plans," and most institutions require them to do so, Martin Conley said. The financial arrangements offered in the plans of various universities, however, are "all over the map," she says. The influence of the stock market on professors' retirement intentions derives in large part from the growing importance of defined contribution plans, under which the employer and employee make set payments into the plan during the person's working life, and payout then depends on the amount of money that has accumulated in the individual's account, which in turn depends on economic conditions and investment strategy. Many public institutions and older private plans, however, offer defined benefits and pay out set amounts determined by such factors as salary and length of service, she explained.

For both the institution and the individual, however, the financial circumstances of potential retirees is only one of many complex issues that complicate the retirement pictures. From the standpoint of university administrators, "the availability of a tenured faculty position essentially as long as you want to hold it, puts institutions in a position where they may or may not be able to make decisions about human resources that private industry has a bit more flexibility to make," Martin Conley said. Several participants wondered whether, for example, well-paid older professors who keep their positions past the traditional retirement age tie up funds that could be used to open tenure-track slots for less expensive younger faculty members.

"Theoretically, [more retirements] would open up positions," Donna Ginther said. In practice, however, universities "are not hiring tenure-track faculty" and are using more and more adjuncts for teaching and soft-money staff scientists or associates for research. In the sciences, noted Committee on Science, Engineering, and Public Policy (COSEPUP) member Linda Abriola, salaries are only a small part of the cost of hiring new tenure-track faculty, with start-up packages and what R. Michael Tanner, vice president and chief academic officer at the Association of Public and Land-Grant Universities, calls "the ever increasing race to give more and more start-up funds to the young stars," a much more significant constraint.

In her 10 years as dean of engineering at Tufts University, said Abriola, she has seen the size of start-up packages rise sharply. "I used to budget $300,000 per start-up package. I am now budgeting $800,000 to a million dollars..." For that reason, "it is not economical for us to bring in too many junior faculty in any given year.... We actually can't afford to have our faculty retire." With about 20 percent of her faculty nearing or at retirement age, "if they all retired at once, we wouldn't be able to teach our classes, and we would be forced to hire part-time faculty."

On the other hand, retirements do tend to move universities toward another important goal, that of diversifying their faculties by race and gender. Younger faculty cohorts are much more diverse than their elders. At the University of California-Berkeley, for example, according to data provided by Goulden, men comprise more than 75 percent of the professors over 70 years of age, but fewer than half of those under 30. Arriving at Princeton in 1978, Joan Girgus recalled, "I was the eighth tenured woman." She also remembered hearing someone say at a meeting that "the best diversity plan a university can have is a good retirement program."

INCENTIVES, NOT COERCION

Institutions that wish to encourage faculty to retire for whatever reason face significant challenges, however, especially making sure that the incentives they use comport with the law. "Two main points of friction [exist] between the legal system and faculty retirement," said attorney Ann Franke, president of Wise Results, LLC, and author of a paper on legal issues regarding retirement for the American Council on Education. One is "whether people are leaving voluntarily or involuntarily" and the other is that "when money is changing hands between an institution and an individual,...there [are] tax consequences or consequences under the nation's pension laws." Institutions must design retirement policies to avoid the reality or even the appearance of coercion or favoritism. They must also examine the goals they wish their retirement policies to meet, design incentives that meet those goals, avoid any disincentives that detract from meeting those goals, and measure their policies' results, she added.

Administrators, including department chairs and deans, must therefore avoid missteps such as stereotyping older faculty or singling out individuals for "targeted discussions of retirement," Franke warned. Learning to avoid such pitfalls may require training in retirement law and proper practices, she said. "Vocabulary missteps," for example, are a major potential snare, as in the case of "a dean of faculty who was advocating to the board of trustees a need for a retirement incentive program," she continued. "His words made it into the *Wall Street Journal*. Here was his argument: 'It is no secret that faculty effectiveness decreases with age and turnover would be healthy. Older faculty members become distanced from the modern roots of their fields.' " His talk described "the yellowed lecture notes, the less-traveled path to conferences and seminars, the less than enthusiastic welcome for students," all of which, Franke said, are "ageist stereotypes."

Administrators are therefore on "thin ice" asking individuals when they plan to retire, she added. "Some lawyers...say you can never ask when people are going to retire." In her view, it is permissible "if you ask everyone in a department, if you ask them in writing, so that the message is consistent. You ask them to put down an age, their name, and you put on the piece of paper that we will not bind you to this. We are doing this for our planning purposes. Then, in my opinion, you are not violating the Age Discrimination in Employment Act."

Another challenge is designing incentives that can motivate so varied and independent-minded a group of individuals. One obvious incentive is money, and the University of California had some success in the 1990s by offering a series of very large incentives, known as the Voluntary Early Retirement Incentive Programs, Goulden said. Princeton is also endeavoring to "incentivize the tenured faculty through bonuses [based on age] to embrace 65 to 70 as an appropriate retirement age," Girgus said. Those who sign an agreement to retire between 65 and 70 become eligible for a one-time bonus of 1.5 times their salary or 1.5 times the average salary of everyone at their rank, whichever is greater.

Princeton has what Girgus calls "a very strong merit salary system," so salaries at each rank vary substantially. Acceptance of the bonus system has been especially high among "those who received a bonus based on the average salary at their rank," she noted. "If you happen to be at the bottom of the salary distribution, the bonus you get is a lot higher than your salary would ordinarily suggest.... It is safe to make the inference that those [receiving the bonus based on the average for their rank] are the faculty that we have judged to be less productive."

REMAINING RELEVANT

Money, however, does not reliably motivate all professors. More than 200 at Berkeley choose to "actually lose income" rather than retire, Goulden says. "Not even taking into account all the private accounts they might have, [counting] just Social Security [and] UC pension," these individuals are losing "$20,000, $30,000, $40,000 [per year] they would get if they were to retire," and one or two "as much as nearly $60,000.... This isn't working for free. They are paying."

This probably happens because work plays a much larger role in the lives of many academics than it does for nonacademic workers, Goulden continued. Only 28 percent of the American workforce says they derive their main satisfaction in life from their work, as opposed to 67 percent of the Berkeley faculty, according to data that he presented. For faculty members aged 65 or older, that percentage rises to 77 percent. One indication of this devotion to work is the long hours that faculty at Berkeley and, presumably, elsewhere put in across their working lives. The average workweek tops 50 hours until faculty members reach age 58, when it dips below that figure for the first time. Not until age 68 does it drop down to "the standard workweek [of] 40 hours," Goulden says.

In fact, the desire to remain connected to the university, the academic community, and the life of the mind that had been their focus for decades can serve as a strong incentive for faculty members to continue working. "I think for many faculty the inhibition to retire is the fear of becoming irrelevant, disconnected, discarded, and forgotten," Tanner said. "The key message that all institutions can do, and not necessarily at great expense, is to say, no, we still value what you have to offer. We want you to be connected. We want to draw on your expertise."

In speaking with senior faculty nearing retirement, interviewers found that many "felt marginalized," said Claire Van Ummersen, senior advisor to the Institutional Leadership Group of the American Council on Education. "They wanted very much to be respected by their institutions, recognized for their contributions, and they were looking for ways to stay connected to the institution that would allow them to be intellectually engaged and have the sense that they were making a difference for that university. These were the most important things that faculty talked about to us."

As an example, John Tully mentioned the story of Nobel laureate John Fenn, who so wanted to remain at work that "he applied for a junior faculty position at his old department." Tully explained that Fenn "had to retire at 70 in 1992 or 1993." After losing out to a younger applicant for the job, "he was threatening...a lawsuit, because he felt that his credentials were every bit as good as the person they hired. Now, that is not as frivolous as it sounds.... He didn't really want a faculty position. He just wanted to be involved in the university."

For this reason, some institutions have developed retirement programs that try to make the separation from regular faculty status gradual, but the connection to the university permanent. One approach growing in popularity is phased retirement, which generally allows faculty members to cut back their work responsibilities gradually, usually over a period of 3 to 5 years. "Touted as a major win-win," this approach has "extended the time that faculty could pay into Social Security,... and shortened the time of complete dependence on retirement savings for income," Martin Conley says. An American Association of University Professors survey done in 2007 showed a rapid increase in such programs, she said. In a survey of 3,300 senior faculty members nearing retirement, 75 percent of respondents preferred phased retirement, Van Ummersen added. At most institutions, individuals phasing out of their careers have left after the third year, even if the agreement allows for 5 years.

Also popular with faculty are retirement systems that have clear guidelines that apply to everyone and are openly discussed and readily accessible. Some institutions negotiate retirement arrangements privately with each individual. Faculty strongly dislike such arrangements, however, because they "didn't know what the rules were," Van Ummersen said. "They didn't know what they could ask for" and feared that others had gotten a better deal. Having very clear policies and guidelines and making the information public clears up this problem.

Many faculty members also wish to continue their scholarly work after retirement, a desire often recognized by emeritus status. They "care deeply, as they are thinking toward retirement, about completing projects that they may be in the middle of," Van Ummersen said. "Am I going to be able to get this finished before I leave? It is that last book that they want to write or some very important research that they want to finish. Anything that the institution can do to help them to do that" will ease the transition, she said.

Princeton, for example, provides emeritus faculty continued access to e-mail addresses, computing privileges, library access, parking permits, and when possible, office space and secretarial and computer support. Those engaged in research may be named senior scholars, with continuing access to research accounts and the ability to accept new postdoctoral fellows (but not graduate students) and apply for certain special grants. On the other hand, City University of New York (CUNY), a very large public institution, doesn't "have any money for the retired professoriate," said Manfred Philipp, the former chair of the University Faculty Senate, the current president of CUNY Academy for the Humanities and Sciences, and a professor of chemistry at Lehman College. "The administrators don't even want to give them e-mail, let alone an office." The faculty union is pressing for the right to retain a university e-mail address after retirement. They are important not only to aid the retirees, he said, but because "retirees still do things like write letters of recommendation for students that they had while they were on the active faculty. They need to have these connections in order to provide these services to students."

Office space for retired faculty is a "real problem," Girgus said, "and lab space, even more. For those faculty that have grants, we try to provide lab space. For those who want offices, we try to provide office space or lockable carrels in the library." Many workshop participants concurred that providing office space for retirees is an often insuperable challenge. COSEPUP member Gordon England suggested that facilities modeled on commercial business centers might meet the needs of some retirees.

An additional approach to keeping faculty connected and active are facilities for retired faculty, either associations or full-fledged emeriti centers. The emeriti center at the University of Southern California (USC), for example, was organized in 1978 at the request of an association that retired faculty had organized in 1949. Despite its name, the center is open to all retired faculty regardless of whether they have the emeritus designation. This "collegial body...does a lot of interesting things," including educational and service projects and social events, said Janette Brown.

The USC center is one of the entities that belong to the Association of Retirement Organizations in Higher Education (AROHE), "a very small nonprofit" without a paid staff that is run as "a labor of love," Brown continued. "Our [AROHE] colleagues are composed of mostly retired faculty, but also leaders on college campuses," she added. "Some of them are from the provost office. Some of them are from HR [human resources], some from benefits, some from the alumni association."

A 2012 survey conducted by AROHE received 117 responses from across the United States and Canada, 75 of them from retiree associations rather than centers. Many institutions lack retiree centers or associations, and those that exist go under many different names, sometimes without a website. The overwhelming majority have fulfillment for retirees as their primary purpose, but many do service or teaching at the university or in the community, and some help prepare faculty members for retirement. They also afford members privileges that can include use of the institution's libraries and e-mail service, use of a university identification card, reduced fees for parking and events on campus, use of the institution's computers, bookstore discounts, and in some cases, use of offices and more.

UNTAPPED ASSETS

Keeping retired faculty connected to the life of the university has important advantages for the institution as well as for individuals, the group agreed. "Your retired faculty and staff, as a body, as a group, are the largest untapped resource that your college or university has," said Brown. Sometimes those advantages can be extremely concrete, noted Philipp, who cited the example of chemist Alan Katrizky, a chemistry professor still active at age 84, who pledged $1.5 million to endow a chair in his field at the University of Florida. In 1980, Katrizky had moved to the Gainesville institution from the University of East Anglia in his native United Kingdom, where he had spent decades as a professor and dean, in order to avoid mandatory retirement at age 65.

Some of what Philipp called the "distinctive assets" of retired faculty are more intangible, but still important. "Strategically engag[ing] vigorous retired professors...can help college[s] and universities maintain a reserve pool of flexible and readily available faculty resources to help institutions adapt to rapidly changing program needs in a time of fiscal constraint." He cited as an example Edward Gerjuoy, an emeritus professor of astronomy and physics at the University of Pittsburgh, who in his mid-90s continues to work in his office 6 hours a day, often including weekends, to write papers, and to do voluntary service for the American Physical Society.

Universities "get so much value out of keeping their former faculty involved," Tanner added. "Many of the former faculty still have a lot of influence externally. They are still working to get awards for their current faculty and writing letters of recommendation and just the wisdom that they can give to the institution. That should be a responsibility of the institution and it will pay off for them."

Fashioning policies that assist faculty in making appropriate choices about when and how to retire, the group agreed, redound to the welfare not only of the individual professors but also of the institution as a whole. "Retirement choice...was, of course, one of the major tenets behind the spirit of the Age Discrimination in Employment Act and the elimination of mandatory retirement," Martin Conley said. "We do still need to think about retirement as a choice, and make sure that we have policies and programs in place that are allowing people to work longer because they want to, rather than to work longer because they have to....What is at stake is nothing short of the quality of life for the academic workforce and their prospects for a comfortable retirement, or perhaps even the ability to retire at all, and in turn, the learning environment for our students."

6
The Other Academe

Presentations at the meeting generally concentrated on academic careers that permit—indeed require—extensive involvement in research. This career path accounts for the overwhelming portion of the scientific discovery and technological innovation emerging from America's academe. It represents, however, only a minority of the jobs open to young scientists and of the faculty members teaching in American higher education. It is also much too small to accommodate the large numbers of Ph.D. scientists graduating from America's universities.

Discussions and presentations at the workshop concentrated on the traditional—and for many people, still the ideal—academic career path at upper-tier research-intensive universities. Frequently appearing in the conversation, however, were considerations of major trends that have led large numbers of Ph.D. scientists who cannot find—or do not wish to compete for—tenure-track positions into the far more plentiful "off-track" and "non-track" jobs in academe.

Trying to "embrace the complexity of the academic career," and specifically of "a scientist career, a science and engineering career," required examining Ph.D.s working in these other academic realms, Valerie Martin Conley said. Edie Goldenberg called attention to the "important" distinction between the academic "haves and have-nots"—a distinction that applies to both individuals and institutions—between the tenure-track and tenured professors at elite, highly endowed research campuses and the educators teaching at community colleges and working in adjunct positions. "These are different worlds," she said. "They are all important, but for different purposes. The goals, the actions that are appropriate have to be, it seems to me, quite different for these different places."

Underlying the importance of community colleges in higher education, Richard Zare, the chair of the Committee on Science, Engineering, and Public Policy (COSEPUP), told the group that 2-year institutions "serve almost half of the undergraduate students in the United States" and perhaps more. In so doing, they "provide open access to postsecondary education, preparing students for transfer to 4-year institutions, providing workforce development and skills, offering noncredit programs ranging from English as a Second Language…to retraining and community enrichment programs or cultural activities."

Community college faculties are almost evenly divided between men and women, he continued. A number of women who teach science, technology, engineering, and mathematics courses at such colleges shared observations on their work in a study funded by the National Science Foundation, Martin Conley reported. Researchers did face-to-face interviews with women at nine community colleges across the state of Ohio and "telephone interviews with women in New York, Florida, Texas, and California, where [there] are large percentages of community college enrollment…. . Over and over again, we were really struck by the responses …from these women, saying, 'I love teaching. I made this choice because I love teaching,' " she explained. "Many of these women were educated at our top research institutions across the country, too. Many of them had Ph.D.s, but the Ph.D. is not necessarily the degree that is required to teach at the community college, so many of them did not."

The women "also talked about their decision not to continue to pursue the Ph.D.," she said. "We really do need to open up our…ideas about what we will accept as that academic career that we are thinking about."

At institutions of every kind, from local community colleges to world-renowned research universities, some members of the faculty work in what Robert Hauser called a "just-in-time, part-time adjunct" labor force. At City University of New York, Manfred Philipp said, "Some of [these teachers] are…doctoral and master's students, but the vast majority come from the academic proletariat, to use a Marxist phrase, that exists in New York City. That is simply the reality."

These teachers "are disproportionately women Ph.D.s with children," the mothers who did not receive tenure or did not secure or even try for a tenure-track position, Mason said. The universities "are feeding this whole labor force in a way that was certainly unintended, but it is occurring. It is a cheap labor force…for the universities to tie into." She added, "This is the fastest-growing part of the academic labor force."

That is because, "if you are sitting in the seat of a provost," Tanner observed, "…adjuncts are a much more effective workforce for educating undergraduates…because they teach more. You have to think carefully about the number of tenured positions you can support."

In addition to spending all their time teaching, adjuncts are economical because they are paid much less than "on-track" faculty for the classes they teach, they generally receive few or no fringe benefits, and they do not tie the university to long-term commitments. For the individual adjunct, that translates to low pay, no job security, and lack of health insurance or an employer's retirement plan. At the City University of New York, "we just got adjunct health care last year," said Philipp. "All of our adjuncts worked without health care. There was some possibility, if they were long-term adjuncts, of getting health care before that....Some adjuncts had the interesting experience of getting sick just before classes start[ed], and then they were dropped from the rolls of the instructional faculty, and then they were dropped from any access to health care, too. This was not an uncommon situation. For adjuncts, getting health care is a difficult sort of situation."

Because of low incomes and lack of employer retirement plans, the same difficult questions raised by the "graying" of the larger population could have an even greater impact on these other academic positions. "Many more people [are] working part-time than we have ever had in history," Martin Conley said. "Both full-time and part-time faculty are aging, as is the population. The average age of part-time faculty has increased from 44 to 50 years old during the time period that they collected the data from NSOPF [National Study of Postsecondary Faculty]." [9]

---

[9] The NSOPF data was collected in four cycles between 1988 and 2004. [SOURCE: https://nces.ed.gov/surveys/nsopf/design.asp]

7
Looking Ahead

"One size does not fit all"—a phrase first enunciated at the workshop by Cathy Trower and then used by various other participants throughout the gathering's two days—emerged as one of the discussion's overriding themes. Universities face a broad range of economic, demographic, academic, and policy challenges. Ever more diverse faculties are dealing with a widening array of personal, family, and professional needs. Ever more unstable financing is affecting different kinds of institutions in different ways. For neither institutions nor individuals, therefore, will a single, broad solution serve to provide resolutions to the issues of the present and the future.

A second major theme is that, given the many changes now under way, the coming years will bring significant changes to some or all institutions. There is not "going to be this continuous line from the past into the future," observed Gordon England. Calling himself an "industry guy," he said that he has seen dramatic disruptions and discontinuities in industrial companies. Given current technologies and trends, "I believe frankly that things are going to change dramatically for universities," he continued. "I really believe universities need to be thinking about it, because it is going to change whether they plan for it or not.... . Education is ripe for dramatic change, because there are so many ways now to disseminate knowledge, different than when we set up universities, however many...hundred years ago."

A major factor changing academe, he suggested, will be "technologies [that] have disrupted a lot of workforces in the last 20 years [but] haven't disrupted the university" yet. But now an academic version of modern information technology—massive open online courses, or MOOCs—has become "one of the hottest topics in education," he noted. These courses will grow in importance with effects on education that are as yet unclear, he said. "People are looking to really reduce their costs, and this might be one of the ways."

But, as Michael Tanner observed, numerous "doomsday scenarios" have been enunciated concerning MOOCs, which he thinks have been "over-interpreted.... They are a new resource, just as Gutenberg's books made possible a number of educational experiences for motivated people.... This new technology is going to open...the library of the world and access to resources that are just unheard of. There will be many more self-educated people out there." But the meaning of these changes for higher education institutions is not yet clear.

Something else likely to drive change at many universities, England predicted, is the explosion in student debt, which he believes will motivate at least some students to make different choices about where or how to pursue their education. "Kids now owe a trillion dollars in student debt," he said, "and that is not going to go up to 2 trillion, so at some point that is going to end." Beyond economics, "the demographics are sort of working against [universities], because the population is getting smaller. The state support is going down [and I think] is not going to improve.... Federal support for research I don't think is going to go up."

One component of the academic scene—the tenure system—that has thus far withstood change remains so important that it constitutes "the elephant in the room," COSEPUP member Paul Citron said. Although it now applies to an ever-shrinking percentage of faculty members, and although in some cases it no longer implies the commitment to provide a faculty member's salary, it nonetheless remains a powerful factor in university finances, organization and culture. "Is it an entitlement whose time has come and gone?" he asked. "Is it something that is an impediment, or is it something that is a major benefit to academia, for whatever reasons?"

Those reasons, participants noted, have historically involved protecting academic freedom. "Tenure has been very important personally in guarding beliefs and allowing faculty to speak out [on] what they see is right, and not be subject to the whims of each year some group saying, 'Well, we can't have someone who thinks that way.' " Richard Zare said.

"I think that tenure is a very essential and important protection for faculty," Edie Goldenberg concurred. "Yes, it is a privilege. Yes, it is costly. Yes, people take advantage of it. The answer in my book is not to get rid of it." Any modifications must still protect "the faculty's ability to go after unpopular things."

On the other hand, John Tully suggested, "I think if we removed tenure it wouldn't change very much, at least if we replaced it with something sensible like a 20 or 25 year contract instead. Whether long-term contracts or other alternatives to tenure could provide sufficient protection merits examination, participants added.

As the academic world sees ever "more pressure on universities to tighten the belts and to watch very carefully where the money is going," the fates of the academic haves and have-nots are likely to diverge even further, Tanner said. At a relative "handful" of prominent, well-endowed research universities, whether private or public, "life won't change a whole lot probably." At many other less-favored institutions, "I think it will change quite a bit."

This means that "for people going along those academic arcs, we owe them very good information about what is happening in the academic marketplace," Tanner continued. To meet changing conditions, he believes, "we are going to have to educate our graduate students to understand a different role.... The students are going to have to be educated to be more flexible, because it is going to be a more rapidly changing world" in which they will need a broader perspective, "as opposed to a very narrow, tightly focused perspective. That tightly focused expertise may not be wanted in the marketplace by the time they get there. I think we do them a disservice if we have not, in fact, given them a sense of alternatives."

Appendixes

APPENDIX A:
WORKSHOP AGENDA

Workshop on the Arc of the Academic Research Career:
Issues and Implications for U.S. Science and Engineering Leadership

September 9, 2013 [8:30 am-5:00 pm] – September 10, 2013 [9:00 am-12:00 pm]
National Academy of Sciences Building – Room 120
2101 Constitution Ave., N.W.
Washington, DC

On September 9-10, 2013, COSEPUP will host a workshop that explores critical stress points in an academic career. The focus will be on career entry, the tenure decision, and retirement. These transition points have been recognized as particularly difficult periods for faculty and administrators. This workshop will bring together experts on the problems associated with each of these career points with the intention of developing a holistic understanding of how policies specific to each area have implications for the others. The workshop will also include presentations about universities that have developed innovative approaches to these problems and discussions by education leaders about opportunities for other actions that could relieve stress on the faculty and improve the efficiency and effectiveness of university operations.

AGENDA:

	September 9, 2013	
8:30 AM	**Chairman's Opening Remarks**	
8:45	**Overview of Challenges to U.S. Universities and Academic Science and Engineering Careers**	
	Shirley Malcom	Director of Education and Human Resources Programs, AAAS
9:30	**The Demographic Context**	
	Robert Hauser	Executive Director, The Division of Behavioral and Social Sciences and Education, National Research Council; Former Director, Center for Demography of Health and Aging at the University of Wisconsin
	Donna Ginther	Professor of Economics, University of Kansas
10:30	**Break**	
10:45	**Getting Started: Early Career Bottleneck**	
	Henry Sauermann	Assistant Professor of Strategic Management, Ernest Scheller, Jr., College of Business, Georgia

		Tech
	Mary Ann Mason	Co-director, Center for Economics & Family Security, University of California-Berkeley School of Law
	Donna Ginther	Professor of Economics, University of Kansas
Noon	Lunch	
1:00 PM	**The Family v. The Workplace: Mid-Career Priorities**	
	Cathy Trower	Research Director, Collaborative on Academic Careers in Higher Education, Harvard School of Education
	Carol Hoffman	Associate Provost and Director of Work/Life program, Columbia University
2:30	Break	
3:00	**Beautiful Sunsets: A Fulfilling Late-career Transition**	
	Marc Goulden	Director of Data Initiatives, Office for Faculty Equity & Welfare, University of California-Berkeley, conducted study of University of California Voluntary Early Retirement Incentive Program
	Ann Franke	President of Wise Results, LLC; author of paper on legal issues regarding retirement for American Council on Education
	Valerie Martin Conley	Professor of Counseling and Higher Education, Co-director, Center for Higher Education, Ohio University, conducted study of retirement for AAUP
4:30	Summary	
5:00	Adjourn	
	September 10, 2013	
9:00 AM	**Reports from the Field: Examples of Innovative Approaches**	
	Joan Girgus	Professor of Psychology, Special Assistant to the Dean of the Faculty on Gender Equity, Princeton University
	Janette Brown	Executive Director of the Emeriti Center, University of Southern California; Executive Director, Association of Retirement Organizations in Higher Education (AROHE)
	Edie Goldenberg	Professor of Political Science, College of Literature, Science and the Arts; Professor of Public Policy, Gerald R. Ford School of Public Policy, University of Michigan
10:30	**Opportunities for Action**	
	R. Michael Tanner	Vice President and Chief Academic Officer, Association of Public and Land-Grant Universities
	Manfred Philipp	Professor of Chemistry, Lehman College; Former Chair, University Faculty Senate; President, CUNY Academy for the Humanities and Sciences

APPENDIX A: WORKSHOP AGENDA

	Claire Van Ummersen	Senior Advisor, Institutional Leadership Group, American Council on Education
	John Tully	Sterling Professor of Chemistry, Professor of Physics and Applied Physics, Yale University
Noon	**Adjourn**	

APPENDIX B:
SPEAKERS' BIOGRAPHIES

JANETTE BROWN, Executive Director of the Emeriti Center, University of Southern California; Executive Director, Association of Retirement Organizations in Higher Education (AROHE)

"Janette C. Brown, Ed.D. serves as executive director of the USC Emeriti Center and Emeriti Center College. In this role, she connects the university with the multigenerational and interdisciplinary cultural and intellectual capital of the USC retiree community, and provides a network of resources, enrichment, and creative opportunities for current and retired faculty and staff and alumni. She works closely with the Center for Work and Family Life and the Office of Benefits Administration to offer wellness programs and resources for healthy aging, retirement transitions and beyond; she also connects students with intergenerational opportunities and internships.

"Dr. Brown also is executive director of the nonprofit Association of Retirement Organizations in Higher Education (AROHE). In her position at AROHE, she has conducted research on issues relevant to retired faculty and staff, including development of the first online instrument to gather extensive data on retiree programs and services at universities in the United States and Canada. She is a member of the National Council on Aging and the American Society on Aging and has presented at programs and conferences in the U.S., Canada, and Germany, and has co-authored a book chapter on the potential of senior scholars and scientists for the European Research Institute on Health and Aging."

SOURCE: http://emeriti.usc.edu/bios/janette-c-brown/

ANN FRANKE, President of Wise Results, LLC, author of a paper on legal issues regarding retirement for the American Council on Education

"Ann Franke has 25 years' experience with national trends in academic policy and education law. She consults nationally with colleges and universities on issues ranging from student injury to academic freedom. She founded her firm Wise Results, LLC, in 2005 after holding senior management positions with United Educators Insurance and the American Association of University Professors.

"Ms. Franke speaks often to national groups, and the American Council on Education invited her to write a book on campus risk management, to appear in 2009. She has published in, among other periodicals, Trusteeship, the Chronicle of Higher Education, Change magazine, Minerva, and The Review of Litigation. She has also served as an expert witness. Ms. Franke is a fellow of the National Association of College and University Attorneys, a trustee of AAUP's Academic Freedom Fund, and a member of the editorial advisory board for "Educator's Guide to Controlling Sexual Harassment.

"Ms. Franke earned her B.A. (magna cum laude), M.A. (linguistics), and J.D. degrees from the University of Pennsylvania and an LL.M. from Georgetown University. Through a Fulbright senior scholar award, she studied the development of private universities in Australia."

SOURCE: http://www2.law.columbia.edu/jfagan/conference/speakers.html

DONNA GINTHER, Professor of Economics, University of Kansas

"Donna Ginther is a Professor of Economics and the Director of the Center for Science Technology & Economic Policy at the Institute for Policy & Social Research at the University of Kansas. Prior to joining the University of Kansas faculty, she was a research economist and associate policy adviser in the regional group of the Research Department of the Federal Reserve Bank of Atlanta from 2000 to 2002, and taught at Washington University from 1997 to 2000 and Southern Methodist University from 1995 to 1997. Her major fields of study are scientific labor markets, gender differences in employment outcomes, wage inequality, scientific entrepreneurship, and children's educational attainments.

"Dr. Ginther has advised the National Academies of Science, the National Institutes of Health, and the Sloan Foundation on the diversity and future of the scientific workforce. She is currently a member of the Board of Trustees of the Southern Economic Association and was formerly on the board of the Committee on the Status of Women in the Economics Profession of the American Economic Association."

SOURCE: http://www.people.ku.edu/~dginther/

JOAN GIRGUS, Professor of Psychology, Special Assistant to the Dean of the Faculty on Gender Equity, Princeton University

"Joan Girgus is Professor of Psychology and Special Assistant to the Dean of the Faculty at Princeton University. She has also served as Chair of the Psychology Department and Dean of the College at Princeton. Prior to going to Princeton, she served as a faculty member and dean at the City College of City University of New York (CUNY). Dr. Girgus has done research and written books and papers on perception and perceptual development, personality development, the transition from childhood to adolescence, and the psychosocial basis of depression. She has also written papers on undergraduate science education and on women in science. Her research has been supported by the National Science Foundation, the National Institute of Mental Health, the National Institute of Child Health and Human Development, the Ford Foundation, and CUNY. Dr. Girgus is one of the principals of The Learning Alliance, the first just-in-time provider of strategic expertise to college and university leaders. From 1993-2003, she was a member of the executive committee of the Pew Higher Education Roundtable and its successor, the Knight Higher Education Roundtable, which worked with a broad range of colleges and universities to identify "best practices" for academic restructuring, and was a consulting editor of Policy Perspectives, which published essays on major issues in higher education. From 1987-1999, she directed the Pew Science Program, a national program to improve undergraduate science education sponsored by the Pew Charitable Trusts. Dr. Girgus is currently a trustee of Adelphi University, the Wenner-Gren Foundation, and McCarter Theatre. She has also served on the Board of Trustees of the American Association on Higher Education (AAHE) and Sarah Lawrence College. Dr. Girgus received her B.A. from Sarah Lawrence College and both her M.A. and Ph.D. from the Graduate Faculty of the New School for Social Research in New York City."

SOURCE: Gender Differences at Critical Transitions in the Careers of Science, Engineering, and Mathematics Faculty, NAP (2010)

EDIE GOLDENBERG, Professor of Political Science, College of Literature, Science and the Arts; Professor of Public Policy, Gerald R. Ford School of Public Policy, University of Michigan

"Edie N. Goldenberg is Professor of Political Science and Public Policy. She served as Dean of the College of Literature, Science and the Arts from 1989-98 and Director of the Ford School from 1987-89. She is the founding Director of the Michigan in Washington Program. Her research interests are in American politics and higher education. Her most recent book, "Off-Track Profs: The Rise of the Teaching Specialist in Higher Education," (with John Cross, MIT Press, 2009) examines the growth in the number of teaching faculty off the tenure track at ten distinguished research universities, identifies the forces driving this trend and the consequences for academic life, and offers recommendations to university leaders for monitoring and managing their faculty workforce. Edie served in the U.S. Office of Personnel Management where she designed and implemented a government-wide evaluation of changes under the Civil Service Reform Act of 1978, including the first systematic survey of the federal workforce. She is a member of the National Academy of Public Administration, a life member of the MIT Corporation, and a member of the Academic Advisory Committee of Yeshiva University where she received an honorary doctorate in 2008 for her contributions to higher education."

SOURCE: http://www.fordschool.umich.edu/faculty/Edie_Goldenberg

MARC GOULDEN, Office for Faculty Equity & Welfare, University of California-Berkeley, conducted a study of the University of California Voluntary Early Retirement Incentive Program

"Marc Goulden, a research analyst at the University of California at Berkeley, has compared the advancement of male and female professors at research universities. For each year after securing a tenure-track job, he found, male assistant professors are 23 percent more likely than their female counterparts to earn tenure."

"[Goulden studied] work/life balance issues with Mary Ann Mason for eight years and presented some new results from their ongoing surveys of

graduate students and faculty members (this research continues to be published in Academe, titled "Do Babies Matter?")."
SOURCE: http://www.zoominfo.com/p/Marc-Goulden/687947270

ROBERT HAUSER, Executive Director, The Division of Behavioral and Social Sciences and Education, National Research Council; Former Director, Center for Demography of Health and Aging at the University of Wisconsin

"Robert M. Hauser has wide-ranging research and teaching interests in aging, social stratification, and social statistics. He collaborated with David L. Featherman on the 1973 Occupational Changes in a Generation Survey, a replication and extension of the classic Blau-Duncan study. Beginning in 1969, he collaborated with William H. Sewell on the Wisconsin Longitudinal Study, and he has led the WLS since 1980. The WLS began as a study of the transition from high school to college or the work force. It has become a multi-disciplinary study of the life course and aging, and the next major round of WLS surveys will begin in mid-2009. In recent years, Hauser has combined work on the WLS with studies of trends and differentials in educational attainment, the role of achievement testing in American society, and the measurement of adult literacy. On these projects, Hauser has worked closely with many graduate students. His classroom teaching repertoire includes social stratification, research methods, and introductory and advanced courses in statistics, including structural equation models and discrete multivariate analysis. He has pursued connections between social science and social policy through his work with the National Research Council."
SOURCE: http://www.ssc.wisc.edu/~hauser/

CAROL HOFFMAN, Associate Provost and Director of Work/Life program, Columbia University

"Carol was recruited to Columbia in 2007 to create the Office of Work/Life. She has launched several new programs at Columbia, including backup care, breastfeeding support, faculty recruitment and relocation service, including spouse/partner dual career service, housing information and referral service, and wellness. Carol has also been able to expand existing programs and policies, such as the School and Child Care Search Service, the affiliated child care centers, flexible work arrangements and faculty family friendly policies.

"Before coming to Columbia, Carol served as the Founding Director of Work/Life at the University of California at Berkeley. Also at UC Berkeley, she founded and directed the employee assistance program, expanded child care opportunities, inaugurated the university's elder care program, and developed programs to respond to trauma, disaster and deaths. She gives presentations at regional and national conferences and has been published on some of these same topic areas.

"Carol is on the Board of the College and University Work Family Association (CUWFA). She received her B.A. from SUNY Buffalo and her

APPENDIX B: SPEAKERS BIOGRAPHIES

M.S.W. from SF State University; Carol is a California-licensed Clinical Social Worker and a native New Yorker."
SOURCE: http://worklife.columbia.edu/node/896/print

SHIRLEY MALCOM, Director of Education and Human Resources Programs, AAAS

"Shirley Malcom is Head of the Directorate for Education and Human Resources Programs of the American Association for the Advancement of Science (AAAS). The directorate includes AAAS programs in education, activities for underrepresented groups, and public understanding of science and technology. Dr. Malcom serves on several boards—including the Heinz Endowments and the H. John Heinz III Center for Science, Economics and the Environment—and is an honorary trustee of the American Museum of Natural History. In 2006 she was named as co-chair (with Leon Lederman) of the National Science Board Commission on 21st Century Education in STEM . She serves as a Regent of Morgan State University and as a trustee of Caltech. In addition, she has chaired a number of national committees addressing education reform and access to scientific and technical education, careers and literacy. Dr. Malcom is a former trustee of the Carnegie Corporation of New York. She is a fellow of the AAAS and the American Academy of Arts and Sciences. She served on the National Science Board, the policymaking body of the National Science Foundation, from 1994 to 1998, and from 1994-2001 served on the President's Committee of Advisors on Science and Technology. Dr. Malcom received her doctorate in ecology from Pennsylvania State University; master's degree in zoology from the University of California, Los Angeles; and bachelor's degree with distinction in zoology from the University of Washington. She also holds 15 honorary degrees. In 2003 Dr. Malcom received the Public Welfare Medal of the National Academy of Sciences, the highest award given by the Academy."
SOURCE: http://www.aaas.org/ScienceTalk/malcom.shtml

VALERIE MARTIN CONLEY, Professor of Counseling and Higher Education, Co-director, Center for Higher Education, Ohio University, conducted a study of retirement for AAUP

"Valerie Martin Conley is Department Chair and Professor of Counseling and Higher Education at Ohio University and Co-Director of the Center for Higher Education. She holds a Ph.D. from Virginia Polytechnic Institute and State University (Virginia Tech) in Educational Leadership and Policy Studies, Higher Education and Student Affairs. She also holds a B.A. and M.A. in Sociology from the University of Virginia. Dr. Conley joined the faculty of Ohio University after an extensive career in institutional research and computer consulting in the Washington, D.C. area, primarily for the National Center for Education Statistics."
SOURCE: http://www.educause.edu/members/valerie-martin-conley

MARY ANN MASON, Co-director, Center for Economics & Family Security, University of California-Berkeley School of Law

"Mary Ann Mason is currently professor and co-director of the Center, Economics & Family Security at the University of California, Berkeley, School of Law.

"Mary Ann Mason's scholarship spans children and family law, policy, and history. Recent works have focused on working families, in particular the issues faced by the surging numbers of professional women in law, medicine, science, and the academic world. Her most recent book (co-authored with her daughter Eve Mason Ekman) is Mothers on the Fast Track: How a New Generation Can Balance Family and Careers (Oxford, 2007).

"From 2000 to 2007, she served as the first woman dean of the Graduate Division at UC Berkeley, with responsibility for nearly 10,000 students in more than 100 graduate programs. During her tenure, she championed diversity in the graduate student population, promoted equity for student parents, and pioneered measures to enhance the career-life balance for all faculty. Her research findings and advocacy have been central to ground-breaking policy initiatives, including the ten-campus "UC Faculty Family Friendly Edge" (http://ucfamilyedge.berkeley.edu/toolkit.html) and the nationwide "Nine Presidents" summit on gender equity at major research universities."

SOURCE: http://www.law.berkeley.edu/3133.htm

MANFRED PHILIPP, Professor of Chemistry, Lehman College; Former Chair, University Faculty Senate; President, CUNY Academy for the Humanities and Sciences

"An ex-officio, nonvoting member of the Board of Trustees and chairperson of the 2006-2007 session of the City University of New York (CUNY) University Faculty Senate, Dr. Philipp is professor and past department chair of chemistry at Lehman College and professor in the biochemistry and chemistry doctoral programs at the CUNY Graduate Center. As a Fulbright scholar in 2005, Dr. Philipp taught bioinformatics and biopharmaceutics at the Catholic University of Portugal.

"He received his doctorate in biochemistry from Northwestern University and his bachelor's degree in chemistry from Michigan Technological University. Dr. Philipp has been program director for the National Institutes of Health (NIH)-supported, research-based student support programs Minority Biomedical Research Support (MBRS), Minority Access to Research Careers (MARC), and the High School Summer Research Apprentice Program. He was co-program director of the NIH-supported Bridges to the Baccalaureate at Bronx Community College and Lehman College. He has also served as national president of the MBRS/MARC Program Directors Organization."

SOURCE: http://www.zoominfo.com/p/Manfred-Philipp/47031059

APPENDIX B: SPEAKERS BIOGRAPHIES

HENRY SAUERMANN, Assistant Professor, Strategic Management, Ernest Scheller, Jr., College of Business, Georgia Tech

"Dr. Henry Sauermann joined the College in 2008. His research focuses on individuals' motives and incentives, and how they interact with organizational and institutional mechanisms in shaping innovative activity. In particular, he studies how scientists' motives and incentives relate to important outcomes such as innovative performance in firms, patenting in academia, or career choices and entrepreneurial intentions. This stream of research also explores important differences in these mechanisms across contexts such as industrial versus academic science or startups versus large established firms.

"In new projects, Dr. Sauermann studies the dynamics of motives and incentives over time, and explores non-traditional innovative institutions such as "Crowd Science" or "Citizen Science" (e.g., https://www.zooniverse.org/). Additional work is underway to gain deeper insights into scientific labor markets and to derive implications for junior scientists, firms, and policy makers."

SOURCE: http://scheller.gatech.edu/directory/faculty/sauermann/

R. MICHAEL TANNER, Vice President and Chief Academic Officer, Association of Public and Land-Grant Universities

"R. Michael Tanner joined the APLU as Vice President for Academic Affairs and Chief Academic Officer in January 2011, where he has led planning for a multi-institutional design project to accelerate development and adoption of "cognitive course wares," which achieve better student learning outcomes in gateway courses. He previously was Provost and Vice Chancellor for Academic Affairs at the University of Illinois at Chicago (UIC) for over eight years, following a 30-year long career at the University of California, Santa Cruz (UCSC). He holds bachelor's, master's and doctoral degrees in electrical engineering from Stanford University. At UIC, he was in charge of 14 academic colleges and the library and had principal responsibility for the budget. He led academic planning and spearheaded major initiatives in interdisciplinary areas, notably a successful NIH Clinical and Translational Science center, and in diversity with an NSF ADVANCE award. At UCSC he was chair of the department of computer and information sciences, acting dean of natural sciences, before becoming academic vice chancellor. He was academic and executive vice chancellor for nine years, serving as the campus's chief operating officer. In 2000, Dr. Tanner was named interim director for the University of California Silicon Valley Center, where he was responsible for planning a satellite campus for 2,000 students at the NASA Research Park, in the heart of Silicon Valley."

SOURCE: http://www.ccst.us/ccstinfo/fellows/bios/tanner.php

CATHY TROWER, Research Director, Collaborative on Academic Careers in Higher Education, Harvard School of Education

"Cathy A. Trower (M.B.A., University of Iowa; Ph.D., University of Maryland), Research Director, has a well-established national reputation as an expert on faculty work/life, including faculty diversity and generational issues, faculty in STEM disciplines and health professions, interdisciplinary work, and general trends in faculty employment. Trower is currently heading up the COACHE Research Institute for scholars interested in using COACHE's robust faculty satisfaction database in their research. She has published numerous articles and several book chapters about faculty work life, and edited a book entitled Policies on Faculty Appointment: Standard Practice and Unusual Arrangements (2000). Prior to coming to Harvard, Cathy was a senior-level administrator of business degree programs, and an adjunct faculty member, at Johns Hopkins University."

SOURCE: http://isites.harvard.edu/icb/icb.do?keyword=coache&pageid=icb.page307143

JOHN TULLY, Sterling Professor of Chemistry, Professor of Physics and Applied Physics, Yale University

"John Tully received a B.S. in Chemistry from Yale University in 1964 and a Ph.D. in Chemical Physics from the University of Chicago in 1968. After two years as an NSF Postdoctoral Fellow at the University of Colorado and Yale University, he joined the technical staff of Bell Laboratories. He served as Head of the Departments of Physical Chemistry (1985-90) and Materials Chemistry (1990-96) before leaving Bell Labs to join the Yale University faculty in 1996. Tully's research centers on the development of theoretical and computational tools aimed at achieving an atomic-level understanding of dynamical processes at surfaces and interfaces, and in the condensed phase. Among the theoretical developments he is pursuing are mixed quantum-classical molecular dynamics, and improved sampling methods for simulation of rare events. He is applying these techniques to examine the rates and pathways of energy flow that accompany the adsorption, desorption or diffusion of a molecule on a surface, as well as to simulate intrinsically quantum mechanical processes such as electron transfer and proton transfer at interfaces."

SOURCE: http://www.crisp.yale.edu/index.php/Faculty

CLAIRE VAN UMMERSEN, Senior Advisor, Institutional Leadership Group, American Council on Education

"Dr. Van Ummersen began her career in higher education at the University of Massachusetts in 1968, serving as an assistant professor of biology, promoted to associate professor with tenure, followed by positions as graduate program director for biology, associate dean for Academic Affairs, CAS, associate vice chancellor for Academic Affairs and interim Chancellor of the Boston campus. Following her tenure at the University of Massachusetts, Dr. Van Ummersen served the new Board of Regents of Higher Education for Massachusetts. As chancellor of the University System of New Hampshire from 1986 to 1992, Dr.

APPENDIX B: SPEAKERS BIOGRAPHIES

Van Ummersen pioneered video links between the USNH campuses to increase teaching efficiency and provide greater access for students to programming. In 1993, Dr. Van Ummersen was appointed president of Cleveland State University. In 2001, Dr. Van Ummersen joined the American Council on Education as Vice President and Director of the Office of Women in Higher Education, a position in which she managed national agendas in support of advancement of women leaders and a system of 50 state networks to identify and develop emerging leaders. In 2005, Dr. Van Ummersen was tasked by ACE to develop a suite of programs to serve higher education administrators from the time they enter administration until they retire as presidents. A new Center was established and Dr. Van Ummersen served as its Vice President for Effective Leadership from 2005 to 2010, overseeing leadership programs for higher education administrators and grant initiatives on higher education issues. Currently, Dr. Van Ummersen serves as Senior Advisor for the Office of Institutional Initiatives at ACE developing customized services for colleges and universities to assist presidents and other campus leaders meet their needs for leadership development throughout the institution. She continues to consult with campuses on issues concerning major challenges they face to be certain that program content stays relevant to their needs."

SOURCE: http://tuftsalumni.org/who-we-are/alumni-recognition/tufts-notables/public-service-education-6/